Do Fish Sleep?

Fascinating Answers to Questions about Fishes

Judith S. Weis

Rutgers University Press

NEW BRUNSWICK, NEW JERSEY, AND LONDON

Library of Congress Cataloging-in-Publication Data

Weis, Judith S., 1941–
    Do fish sleep? : fascinating answers to questions about fishes /
Judith S. Weis.
        p. cm. — (Animal Q & A)
    Includes bibliographical references and index.
    ISBN 978-0-8135-4941-5 (pbk. : alk. paper)
    1. Fishes—Miscellanea. I. Title.
    QL617.W36 2011
    597—dc22
                                                        2010017320

A British Cataloging-in-Publication record for this book is available
from the British Library.

Visit our Web site: http://rutgerspress.rutgers.edu

Manufactured in the United States of America

Contents

THREE Fish Bodies 25

FOUR Fish Lives 49

Do Fish Sleep?

Some Major Orders of Teleost Fishes

Elopiformes—tarpons and bonefish
Anguilliformes—true eels
Clupeiformes—herrings, anchovies
Salmoniformes—salmon, trout, smelt, pikes
Cypriniformes—minnows, carps
Siluriformes—catfishes (closely related to Cypriniformes)
Gadiformes—cod, haddock, whiting
Beloniformes—flyingfishes, needlefishes (related to
    Cyprinodontiformes)
Cyprinodontiformes—killifishes, topminnows, guppies,
    mollies
Atheriniformes—silversides
Gasterosteiformes—sticklebacks, seahorse, pipefish
Perciformes—perch, barracuda, gouramis, basses, sunfish,
    cichlids, cardinalfishes, porgies, goatfishes, damselfishes,
    butterflyfishes, angelfishes, grunts, blennies, jacks, drums,
    wrasses, parrotfish, gobies, swordfish, tuna, mackerel
Scorpaeniformes—scorpionfishes, rockfishes, searobins,
    stonefishes
Pleuronectiformes—flatfishes: flounders, soles
Tetraodontiformes—puffers, boxfishes, triggerfishes, file-
    fishes, ocean sunfish
Lophiiformes—anglerfishes, frogfishes

have bones inside like limbs of land vertebrates. Thought extinct for a long time, they have a living representative, the Coelacanth (See this chapter, question 5: Which fish is referred to as a "living fossil"?).

Dipneusti are the lungfishes, with six freshwater species in tropical Africa, South America, and Australia.

Actinopterygii are the ray-finned fishes, the dominant group of living fishes. Their fins are flat with thin bony rays for support. Within this group are the following:

Chondrostei, the more ancestral ray-finned, secondarily cartilaginous fishes, sturgeons and paddlefish, all fresh water.

Holostei, the intermediate ray-finned fishes, gars and the bowfin. Only eight species remain, all in North America.

Teleostei, the advanced bony fishes, the most diverse and successful group with 425 families. Teleosts are found in all oceans and most freshwater areas, from the abyss to the estuaries, from mountain streams to thick swamps, from hot springs to the freezing Arctic. Their modes of life and behavioral and anatomic adaptations are diverse and fantastic. Their evolution has generally been from soft-rayed fishes with generalized anatomy to more specialized fishes with spiny, rayed fins. The 23,600 currently named species of teleosts comprise over half of all species of living vertebrates and over 96 percent of all living fish species. They will receive the most attention in this book, for obvious reasons.

Question 5: Which fish is referred to as a "living fossil"?

Answer: In 1938, Marjorie Courtenay-Latimer, the curator of a local museum, found a most unusual fish in East London, South Africa. She was looking through a trawler's catch and found a 5-foot-long, 127-pound specimen with fins that looked like limbs. She informed prominent South African ichthyologist J.L.B. Smith of Rhodes University about it and sent him a sketch of the fish. By the time he was able to observe the specimen, it was quite decomposed, but he recognized it as a coelacanth, a crossopterygian fish.

The discovery created a worldwide sensation because the ancient crossopterygians were thought to have all disappeared sixty-five or seventy million years ago. Crossopterygian fishes lived before land vertebrates evolved and two hundred million years before the dinosaurs! They had thick, lobed fins, the fore-runners of arms and legs, with the same bone structure as limbs of land vertebrates.

The coelacanth appears to be a cousin of the fish credited with first growing legs and coming ashore—360 million years ago—as the ancestor of all land vertebrates. Smith named the fish *Latimeria chalumnae* after Courtenay-Latimer and the Chalumna River near where it was caught. Eager to find more specimens, scientists searched for years in the Indian Ocean along the east coast of Africa. In 1952, a second specimen was obtained near the Comoros Islands. Additional specimens were subsequently collected off the coasts of South Africa, Tanzania, Kenya, and Madagascar, generally at depths of over three hundred feet.

The scientific community was amazed again in 1998 when Mark Erdmann, a University of California, Berkeley, researcher, discovered a coelacanth in North Sulawesi, Indonesia, in the Pacific Ocean—far from the Comoros Islands. Erdmann was on his honeymoon when he spotted one in a fish market. He then interviewed fisherman around North Sulawesi and found two men who said they occasionally caught the coelacanth, which they called *raja laut,* or "king of the sea." After carefully monitoring the catch for several months, scientists identified a second Sulawesi coelacanth from deep waters (over 300 feet) in Bunaken Marine Park. The Sulawesi fish is brownish-gray rather than bluish, like the Comoran fish. Detailed molecular and anatomical studies revealed that the Comoran and Sulawesi fish are different species that diverged several million years ago. The Sulawesi fish has been named *Latimeria menadoensis.*

Coelacanths reach up to 6 feet long (fig. 2), and preserved specimens can be found in museums around the world. They seldom survive capture and hauling to the surface, though they have been videotaped in their deep, rocky habitat in Sulawesi by remotely operated vehicles (ROVs). They are very slow swimmers that alternate movements of their two pectoral fins. Coelacanths have unique patterns of spots that enable researchers to identify individuals, which tend to stay in the same area for long periods of time. In the 1970s, scientists at the American

Figure 2. Coelacanth, *Latimeria chalumnae. (Photo by Alberto Fernandez Fernandez, Wikimedia.)*

Museum of Natural History, investigating the internal anatomy of their specimen, found it was a pregnant female carrying several foot-long, well-developed embryos, another amazing discovery about these fish. There are undoubtedly many more fascinating things to learn from these "living fossils."

Question 6: How many species of fishes exist?

Answer: Over twenty-five thousand species have been named and identified, but each year many new ones are discovered, so the number continues to increase, despite those that have been going extinct. Most of the new species are being found in the sea because new equipment allows us to explore more of the vast and remote deep-sea areas. Scientists working on a huge census of the world's oceans have discovered about five thousand previously unknown but not yet all named fish species.

Question 7: How long does a fish live?

Answer: Fish longevity varies considerably. Some small fishes live less than 1 year and some very large ones can live for many decades. Sturgeons, for example, live over 50 years. Of course, if taken care of properly, fish can live far longer in captivity, with food provided and no predators around. An Australian lungfish, *Neoceratodus forsteri*, at Chicago's Shedd Aquarium has been there for 75 years and may be over 80 years old. Koi (domesticated varieties of the common carp) in backyard ponds can live for around 70 years. According to a koi seller, the oldest recorded koi was passed down by generations of a Japanese family and lived for 276 years. Many popular aquarium fishes can live a decade.

The fish with the shortest known lifespan, the coral reef pygmy goby, *Eviota sigillata*, lives only two months. It has the shortest life span of any vertebrate known. According to Martin Depczynski and David Bellwood of James Cook University, the tiny fish spend their first three weeks as larvae in the open ocean, then settle on the reef, mature in two weeks, and have an adult lifespan of just three and a half weeks.

Question 8: How can you determine the age of a fish?

Answer: Fortunately for us, individual teleosts keep a permanent record of their life history in their bodies' hard tissues. We can tell the age of a bony fish by examining its cycloid or ctenoid scales (see chapter 3, question 3: What are scales for?) because they grow as the fish grows. As the scale grows, it lays down concentric circular bands called *circuli*. Large spaces between circuli indicate favorable environmental conditions, plentiful food, and fast growth. During periods of environmental stress or decreased metabolism, such as winter, growth is slower, so the circuli form very close together and can be so close that they appear as a single heavy line on the scale. These heavy lines, or *annuli* (year marks), like growth rings of trees, indicate periods of growth and rest. Each winter these fish produce an annulus, and by examining a scale under a microscope, you can count annuli to determine the animal's age. However, the separation between annuli may become blurred due to the partial reabsorption of the outer edge of scales or be unclear in older fishes that are not growing rapidly even in the summer. If a scale is lost or removed, a new one will replace it, but it won't have the history of previous annuli on it. A biologist typically takes scales

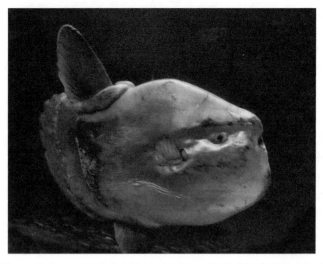

Figure 3. Ocean sunfish, *Mola mola*. *(Photo by Fred Hsu, Wikimedia.)*

from behind the pectoral fin, where they are protected and less likely to be lost and replaced. The same pattern of annuli can be seen in cross sections of bones and in the ear stones (otoliths), which, because they are usually more accurate than scales, are used when more precision is required. However, examining otoliths is more involved than examining scales and requires special equipment to slice and polish them for study.

Question 9: Which is the smallest fish? The largest?

Answer: Fish range in size from the 51-foot whale shark *Rhincodon typus* (see color plate A) to a less than one-half-inch fish in the minnow family—the tiny *Paedocypris progenetica*, discovered in Sumatra in 2006. It is the world's smallest vertebrate (as of 2009) and lives in acidic water that would kill most other fishes. The largest teleosts include marlins, which can exceed 11 feet and weigh 1,200 pounds. The ocean sunfish, *Mola mola* (fig. 3), reaches at least 10 feet and can weigh over 1,000 pounds. Oarfish are large and greatly elongated but thin. One species, swimming. They must tolerate large changes in salinity and temperature as the tides go up and down. Sandy beaches are not as rich in food as muddy areas, but sculpins, flounders, soles, surfperches, mullets, and silversides will move up with the tides. Pacific coast grunions migrate up sandy beaches to spawn and deposit their eggs during high tides. Below the low-tide line and out to the continental shelf, the benthic environment is more stable and a rich area for fish. Many additional fishes live in the varied bottom habitats here, including skates, sharks, eels, cods, pipefishes, drums, wrasses, and triggerfishes. Kelp forests and sea grass beds are particularly productive habitats and house many juvenile fishes. Rocky bottoms support many fishes including rockfishes, *Sebastes* spp. (see color plate B), redfish, and red snapper.

Deeper water is less productive and provides fewer different habitats, so there are fewer species near the bottom, but flatfishes, rockfish, and cods are found there. Below 1,000 meters, light becomes dim. This zone depends on food falling down from above. Some of the same families found in shallower water can live here, but other species like chimaeras, rattails, hagfish, and electric rays are only found in deep water. The lower slope and abyssal zone that may be over a mile deep have uniform darkness and temperature year-round, slow currents, high pressure, and less food. Fish here are scarce and small. Some have elongated rays on fins that prop them up on the soft bottom. One of the most bizarre-looking fish is the blobfish, *Psychrolutes marcidus,* that lives in the deep sea off the coast of Australia. It withstands the high pressure of these depths because its body is soft and very gelatinous, with a density just less than water. This allows it to float right above the bottom without using much energy.

Pelagic fishes live in a more uniform, open water environment. They tend to be strong swimmers and feed on plankton or other fishes. The upper water (*epipelagic* zone) is the warmest and best lit, but is subject to daily fluctuations in temperature and light. It contains most of the photosynthetic phytoplankton, the most important primary producers in the sea, so it has a lot of animal

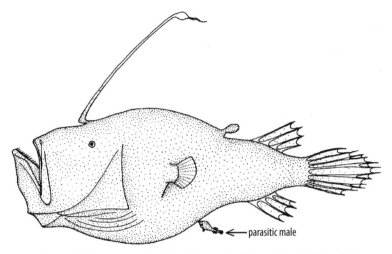

parasitic male

Figure 5. Deep-sea anglerfish. *(Drawing by Tony Ayling, Wikimedia.)*

life as well, including fishes. About seventy families of fishes live here, including strong-swimming carnivores like tuna, shark, and swordfish, the weak-swimming and floating ocean sunfish (see fig. 3, in chapter 1) and a variety of smaller fishes.

In deeper waters (*mesopelagic* zone), the light dims and phytoplankton diminish. This zone includes lanternfishes and other bioluminescent fishes (see color plate B, and see chapter 3, question 17: How do some fishes produce light?). Mesopelagic fishes tend to have large eyes and mouths. Some of them make vertical migrations to the surface at night to feed. In the deep sea or *bathypelagic* zone, it is always cold and dark and food is scarce. Fish are rare and small but have huge mouths and distensible stomachs, like gulpers and anglerfish, to take advantage of the occasional prey that does come along (fig. 5).

Question 6: Which fishes are found in hot water and how do they survive?

Answer: While many species inhabit warm water, Death Valley desert pupfish, *Cyprinodon salinus,* live in isolated ponds with

THREE

Fish Bodies

Question 1: How do fishes swim?

Answer: The study of fish swimming encompasses many aspects of biology: biomechanics, physiology, ecology, and behavior. The yellow perch, *Perca flavescens,* is a common game fish that travels in schools and generally prefers shallow waters near the lakeshore. It swims like typical fishes—its spine curves from side to side somewhat like a snake slithering along the ground, and the tail propels it forward as it bends in the opposite direction from the rest of the body forming a slight S-shape. The curving is accomplished by contracting and relaxing a succession of muscle blocks, called *myomeres,* alternately on each side of the body, starting at the head and progressing down toward the tail. The alternate shortening and relaxing of successive myomeres, bending part of the body first toward one side and then toward the other, results in a series of waves traveling along the fish's body. The end of each wave thrusts the tail against the water and propels the fish forward. Eels and other snake-like fish undulate in this fashion through the water, like a snake. This is relatively slow and requires a lot of energy. Because movement of the head back and forth produces drag (consuming additional energy and slowing speed), many fishes have modified this motion by keeping the waves very small along most of their body and then increasing them in the tail region. The end of the traveling wave moves the tail forcefully back and forth, providing the main propulsion for forward motion. The speed that fish can obtain depends on their size, shape, swimming style, and other factors. Generally, the maximum speed of a small fish is about ten times the body length per second. Speed increases with size and with the frequency of the tail beat. There are great differences between what fishes can do in a sudden spurt when frightened (necessary for survival) and what they can do in a sustained manner—the difference between sprinting and running a marathon. Fins act as stabilizers to prevent the fish from rolling and as brakes to assist with difficult maneuvers. This is important in rocky or coral areas where fishes need to chase prey around corners or make short, sharp turns.

The density of water, about eight hundred times that of air, dictates that bodies moving through it be streamlined, as designers of submarines know. The principles of streamlining are obvious in the bodies of fishes that are fast swimmers. Drag is minimized by the streamlined shape and by a slime they secrete from their skin that maintains laminar (smooth) flow of water. It is not unusual to find speeds around 20 to 30 miles per hour. Tunas and swordfish can swim even faster and have the most streamlined shape, the smallest undulations of the body, and fold up fins that would otherwise stick out into the water and create drag to slow them down. Their tails appear to provide the greatest thrust. Tunas may swim 50 miles per hour; and swordfish, 60 miles per hour.

Not all fishes swim this way. Inflexible-bodied fishes like trunkfish simply alternate contractions of the muscles on each side of the tail, causing the tail to wag from side to side. Some species use fins to propel themselves but none of them swim particularly fast.

Question 2: What are the different functions served by fins?

Answer: Fins are appendages made up of bone and connective tissue that fishes use to move, steer, and stop. There are single fins along the midline, including the *dorsal* fin on the back, marine fishes still need some buoyancy mechanism unless they spend their lives on the bottom. Buoyancy can be increased by storing oil in the tissues (oil being lighter than water) or by having air or gas inside the body. Sharks and rays store lightweight oils in their livers to make them more buoyant. Also, as a shark swims, its pectoral fins provide lift, much the same as the wings of an airplane. Despite the oil, sharks are still denser than water and must swim constantly or sink. A better solution is to be able to adjust the density of the body to match that of the water.

The milkfish, *Chanos chanos*, is a typical teleost. Found in the waters of Southeast Asia and an unofficial national symbol of the Philippines, the toothless milkfish swims in schools and eats algae and small invertebrates. It maintains its buoyancy much like any other teleost—through a specialized, gas-filled organ called a swim bladder or gas bladder, which is evolutionarily related to lungs. It allows them to be neutral at whatever depth they are swimming, without tending to sink or rise up. They regulate the amount of gas in the bladder with a gland that can either secrete or absorb gas; when it is more inflated they are more buoyant and higher in the water, and when there is less gas they sink down deeper, like a scuba diver adjusting the amount of air in the buoyancy-control vest in order to swim effortlessly at different depths.

There are two general types of swim bladders, open and closed. The open swim bladder connects to the esophagus (between the throat and the stomach), giving fish (e.g., eels and salmon) the ability to gulp in or belch out air to adjust their buoyancy quickly. Those with a closed swim bladder, (e.g., cod) lack this connection and must rely only on the gas gland. The combination of gases in the bladder varies; in shallow water fishes it is close to that of the atmosphere, while deep-sea fishes tend to have more oxygen.

If the swim bladder is not functioning properly, a fish may float to the surface involuntarily and have to struggle to swim downward, or vise versa. The fish may lean to one side, float upside down, swim with its head pointing downwards or upwards. These problems indicate difficulty in removing gas from or adding gas to the bladder. When fishes are brought up too quickly from very deep water, the swim bladder expands so rapidly that it might explode or protrude from their mouth. While the same physical principle is involved (gases expanding as pressure decreases), this is different from "the bends" that inflicts scuba divers who ascend too rapidly. The bends happens when gases dissolved in the blood bubble out of solution as the pressure decreases, damaging capillaries throughout the body.

Some fishes do not have swim bladders. These include the more primitive sharks and rays as well as teleosts like flounders that live at the bottom and do not make major excursions up into the water column.

Question 6: How do fishes breathe under water?

Answer: Breathing is essential—all living things must take in oxygen in order to respire and also must eliminate carbon dioxide, respiration's waste product. Fishes use gills to extract oxygen from the water. Since water contains relatively little oxygen (far less than air), they must be able to obtain oxygen efficiently. To exchange gases (dissolved oxygen from the water moves into the fish and carbon dioxide from the fish moves into the water), blood and water are brought into close contact on different sides of a thin membrane, usually only two cells thick, through which the gases diffuse. A gill is composed of many thin-walled filaments containing capillaries that the blood moves through, which allow carbon dioxide to move out of the blood (where it is more concentrated) into the water and oxygen to go from the water (where it is more concentrated) into the blood. The filaments are branched and attached to bony arches. Filaments, being numerous and branched, have a lot of surface area for exchanging gases. Active fishes like tuna need more oxygen and so have more gill filaments and surface area than sluggish fishes that spend their time inactive on the bottom. The effectiveness of the gills depends on the rate at which water can move through. Fishes actively pump water by taking it in through their mouths and pumping it through the gills. First the mouth opens and the

Figure 9. Spookfish, *Dolichopteryx longipes. (Photo by Tamara Frank.)*

spookfish (fig. 9) appears to have four eyes, but has the usual two, each of which is split into two parts: one half points upward to capture faint light from above, and the other half, which looks like a bump on the side of the head, points downward and uses a mirror to reflect the occasional light from bioluminescence onto the retina. The "barreleye" fish has odd tubular eyes, capped by bright green lenses that point up, and a transparent head. Until recently, their eyes were believed to be fixed in place and to provide a silhouette of only whatever was directly above their head. Bruce Robison and Kim Reisenbichler of the Monterey Bay Aquarium Institute made video observations that show that the eyes can rotate within the transparent shield, allowing the fish to peer up at potential prey or focus forward to see where it is going.

Question 9: Do fishes smell? (Do fishes detect odors?)

Answer: Fish have an amazingly keen sense of smell (olfaction) for detecting chemical information in the water. Their olfactory organ consists of two openings, or nares, above the snout that resemble nostrils. Nares open up into a chamber lined with sensory pads that perceive chemicals in the water and send the information to the brain. The ability to move water rapidly over these sensory pads is essential for a keen sense of smell. Fishes use olfaction to locate food, detect predators, recognize kin, navigate, and find a mate. For those that live in dark caves or at great depths, these organs are particularly sensitive.

Odor can also be used to signal alarm. Many fishes, especially freshwater minnows, react with alarm to substances produced by an injured member of their own species. The injured or frightened fish secretes minute amounts of alarm pheromones from special club cells in its skin. This conveys information that a predator is nearby, causing the others to move away.

Odor can also be involved in navigation and migration. Salmon (see color plate C) make extensive migrations from streams where they are born down to the ocean, where they live for many years. When they are ready to reproduce, they find their way up the same river and eventually back to the original stream where they were born—an amazing feat of navigation for someone without a road map. About fifty years ago, scientists studying this remarkable homing behavior found that it was due to the sense of smell. The young salmon become imprinted and form critical "olfactory memories" before leaving their home stream, and these memories eventually guide them back home from the sea to the stream where they hatched years before.

Recently it was found that larval coral reef fish of Australia's Great Barrier Reef also find their way back to their home reef across miles of open sea by olfaction. The tiny larval fish spend their early days swept up in ocean currents that disperse them far from their place of birth. Individual reefs contain different fish populations, and Gabi Gerlach of the Marine Biological Laboratory and colleagues were intrigued at how tiny damsel and cardinal fishes, born on one reef, managed to find their way back home, braving strong currents and predators during up to twenty days at sea. Like salmon, they have olfactory imprinting and can discriminate specific odors in ocean currents. They seem to detect the scent of their natal reef and follow it home.

Question 10: Can a fish hear?

Answer: For humans, the sea is an almost silent world since our ears are virtually incapable of hearing under water. Because water is much denser than air, more energy is required to produce sound in water, but when it is produced, it travels faster than in air and does not dissipate as rapidly. Sound can be produced by earthquakes and volcanoes, by animals, including fish and marine mammals, and by ships and industrial operations. Although fishes do not have external ears or middle ears like humans, they have inner ears that are sensitive to sound waves. The common goldfish, *Carassius auratus,* for example, is very sensitive to noise, and if subjected to loud noise for long periods of time will suffer stress and lose hearing ability. (So don't play loud rock music in the same room with your goldfish!) Their inner ear, like ours, has bony structures (*otoliths*) that respond to gravity and are used for both sound detection and equilibrium. The ear has three fluid-filled semicircular canals, each of which contains an otolith suspended above a membrane covered with over one hundred thousand hair cells (similar to the cochlea in our ears). Sound oscillates the otolith with respect to its surroundings, stimulating the sensory hair cells that send impulses to the brain—the essence of the sense of hearing. In other words, the otoliths hang in their fluid chamber, sound waves cause the fluid to oscillate, and, since the otoliths are denser than the fluid, they move more slowly and touch the sensitive hairs. It is interesting to note that the shape of the otoliths is different for nearly every species of fish, and scientists can often identify a species just from the otoliths. It is possible to examine a predator's stomach contents and determine what fish it ate by the otoliths found there.

Fishes can respond to a wide variety of frequencies, and those with the keenest sense of hearing have a connection between the ear and the swim bladder (see this chapter, question 5: How do fishes prevent themselves from sinking?). This connection can be a series of small bones (extensions from vertebrae, called

Webberian ossicles) from the swim bladder to the ear. The swim bladder resonates in response to sound and magnifies it, and the bones transfer the vibrations to the ear. This type of connection is found in minnows and catfishes. Within these groups, goldfish can detect 40 to 3,200 Hz (about the range of a piano) and brown bullheads hear 100 to 4,000 Hz. Herrings have a pair of elongated gas ducts rather than bones connecting the swim bladder to the inner ear. Fishes without any connection respond to a much lower range of sounds. For example, Atlantic salmon can hear only 40 to 350 Hz. However, David Mann and colleagues at the University of Maryland found that the American shad can detect ultrasound, inaudible to our ears, of 25,000 to 130,000 Hz, which may be an adaptation to detect ultrasonic clicks of their dolphin predators.

Question 11: Does a fish have a sense of taste?

Answer: They certainly do, and their taste receptors are not only inside the mouth but also on the lips, head, and other parts of the body. Fish with barbels (that may look like whiskers) have dense taste buds there. Catfish, carp, and other whiskered fish like goatfish drag their barbels along the bottom to find food. Bullhead catfish have over a million taste buds all over the body, but mostly on the barbels. In addition to taste buds, fish may have single-celled taste receptors called spindle cells scattered in the skin, which may outnumber cells in taste buds.

The taste buds on different parts of the body have different capabilities and thresholds. While the sense of smell is important in finding food, taste is important in the eventual selection and ingestion of the food. By having taste receptors outside the mouth, the fishes should take in only what they like—there should be no need to spit out something that turns out to taste bad. Yet sometimes fishes do take food into their mouths then spit it out. Perhaps the taste buds inside the mouth detect something that the external taste buds do not.

In at least one fish, the sense of taste is used in navigation. John Bardach and his colleagues at the University of Michigan found that catfish in the genus *Ictalurus* can find distant chemical clues by means of taste alone. Removing the sense of smell did not impair their searching ability, but unilateral (one-sided) removal of taste receptors, which are spread over body, and barbels caused them to circle toward the intact side.

Question 12: Are fishes sensitive to touch?

Answer: As in most animals, the skin of a fish is endowed with sensory nerves; many of these nerves relate to the sense of touch, while others react to changes in pressure or temperature. This sense of touch is distinct from the lateral line (see next question) and is most important in species that live in close association with objects—for example, sea robins on the sea floor, coral reef fish, and those that associate with aquatic plants. For some, the sense of touch is also important during mating.

There has been considerable controversy over whether fishes feel pain. A group of scientists led by Lynne Sneddon of Roslin Institute found sites in the brains of rainbow trout that responded to damaging stimuli. They investigated the fish for the presence of *nociceptors,* sites that respond to tissue-damaging stimuli, and found fifty-eight receptors on the face and head that responded to at least one of the stimuli, of which twenty-two could be classified as nociceptors. The presence of nociceptors is not enough to demonstrate that fish feel pain, so they injected bee venom or acetic acid into the lips of some of the trout, with control groups receiving saline solution injections or simply being handled. Trout subjected to bee venom and acetic acid altered their behavior and demonstrated a rocking motion, strikingly similar to the motion seen in stressed mammals. Venom-injected trout also rubbed their lips onto the gravel in their tank and on the tank walls, which did not appear to be reflex responses. Injected fish also took almost three times longer to resume feeding than the controls. While the investigators interpreted their findings as fish feeling pain, others, including angling groups, dispute this interpretation. The findings are relevant to catch-and-release fisheries, in which studies find some injury, altered behavior, or physiology, suggesting that there is some injury and stress from being hooked.

Question 13: What is the lateral line organ?

Answer: It is hard to imagine a type of sense that is totally foreign to us, but fishes possess one, the lateral line. The lateral line system is related to the ear, and together they are called the *acoustico-lateralis* system (fig. 10). However, while the ear responds to sound waves, the lateral line responds to water movements. The roach, *Rutilus rutilus,* is a common freshwater fish in Europe and Asia that reaches about 1 foot long. It is often found in schools in moving water, 6 to 10 feet deep. It uses the lateral line to aid in schooling. Its lateral lines are visible running along each side, from the head to the base of the tail. Within the canals are a series of small organs (neuromasts), each with hair cells located inside a gelatinous capsule, or cupola. The cilia of the hair cells are bent by water movements, and this information is sent to neuromasts, which convey it to the brain. The hair cells are generally oriented so that they respond to currents moving along the fish's body, and adjacent hair cells are generally oriented in different directions. The hair cells and cupulas may be at the bottom of a pit or groove. If a lateral line organ is covered by a scale, the scale will have a pore so that water movement can be detected.

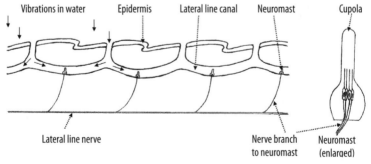

Figure 10.   Lateral line. *(Drawing by Peddrick Weis.)*

a luminescent "lure"), for communication, for species recognition, for courtship, or as a warning that a predator is near.

Question 18: How do some fishes produce sound?

Answer: Most people are unaware that fish produce sound because we usually can't hear it. But when hydrophones are placed in the water, it turns out to be a very noisy place. A number of fish regularly make noise, primarily with their swim bladders. These hollow structures can resonate like a drum. One of the noisy fish actually is called a "drum." Others are named "grunts" or "croakers" because of their sounds. The swim bladder can produce sound several ways. The most common way to is to vibrate it with special (sonic) muscles on the body wall. The Gulf toadfish, *Opsanus beta,* contracts its sonic muscles (the fastest-known muscle of any vertebrate) against its swim bladder thousands of times a minute to generate a loud drone. Some catfish strike the bladder with a springy apparatus, and cusk eels rattle bones against it. Herring release air bubbles from the swim bladder through their anus in a "fast repetitive tick" (you can figure out the acronym).

Fish can also produce sound by rubbing skeletal parts together, *stridulation,* which produces grinding or snapping noises. For example, some marine catfish have specialized pectoral fin spines that make a stridulatory squeaking sound. The northern seahorse produces sounds using the bony edges of the skull, which produce snaps and clicks. Stridulatory sounds may be amplified by the swim bladder.

Sound production can be used as a warning or alarm signal when predators approach. Other sounds are associated with mating behavior. Male toadfishes, *Opsanus,* make boat whistles, grunts, and growls to attract females. Given the appearance of toadfish (see color plate D) it is no wonder they use sound to attract a mate.

In a town in Florida, citizens who had been distressed by hearing eerie thumps were shocked to discover that the noise was being produced by fish. The black drum emits mating calls that travel at a low enough frequency and have a long enough wavelength to carry through sea walls, into the ground, and through the walls of waterfront homes. Noisy fishes near Block Island, Rhode Island, turned out to be the tiny striped cusk eel, which can sound like a jackhammer.

Figure 13. Piranha, food's eye view.
(Photo by Kos, Wikimedia.)

the South American piranha, (actually several species in four genera) hunts in packs and is equipped with sharp teeth suitable for cutting pieces of flesh and tearing them from the prey (fig. 13). Although they are only 6 to 8 inches long, groups of piranha can kill a large animal.

Other fishes use a very different strategy—ambush. They sit in one spot and wait for the food to come to them. These are often inconspicuous and camouflaged, like scorpionfish (see color plate H), and may regularly stake out a particular area for a feeding station. Anglerfishes (see fig. 5, in chapter 2) increase the chances of prey coming near by wiggling a lure.

Question 3: Do fishes migrate?

Answer: All fishes move around to some extent, and some move extensively enough that it can be called migration. Migration to a new location may be in search of food, warmer water, or breeding territory. Short-distance feeding migrations are made by many fish, including vertical migrations (*diurnal vertical migration*) from darker, deeper waters up toward the surface at night to feed while protected from daytime predators.

Tunas make major migrations hundreds of miles along the coasts and across the ocean. Some move across the Pacific

Ocean from California to Japan and back in the course of a year. We have learned a great deal about the movements of large, highly migratory pelagic (open water) fishes as a result of radio-tagging by Barbara Block of Stanford University. She and her colleagues use a new type of remote telemetry device, called *pop-up satellite archival tags,* which are essentially computers that record navigational information, body temperature, depth, and water temperature. Information gained from these tags is improving our understanding of the biology of these commercially and ecologically important species, and it is critical to their future management and conservation.

While it was previously thought that all manta rays stayed in shallow waters, only the recently distinguished smaller mantas, newly named *Manta alfredi,* do that. Andrea Marshall of the Manta Ray and Whale Shark Research Center in Mozambique found that the larger giant species, *Manta birostris,* dives deeply, and migrates 700 miles in just sixty days from Mozambique to the Maldives, the longest migration known for a fish living in the Indian Ocean. In the Maldives, there is a site where hundreds of the fish gather, apparently for breeding.

Every year in June and July sardines, the pilchard *Sardinops sagax,* spawn in the cool waters of the Agulhas Bank and move northward along the east coast of South Africa to warmer waters. The "sardine run," containing millions of individuals, attracts large numbers of predators including sharks and dolphins from below and gannets from above. It is not unusual for the mass of fish to be four miles long and a mile wide, clearly visible from the surface of the ocean. Sardines group tightly together when threatened, forming "bait balls" 50 feet in diameter and extending to a depth of 30 feet (fig. 14).

Some of the most extensive migrations are between sea water and fresh water, termed *diadromous* behavior. *Anadromous* fishes like salmon spend most of their lives in the ocean and move to fresh water to breed, while *catadromous* fishes like eels, *Anguilla,* spend most of their lives in fresh water and move to the ocean to breed. Prior to migrating, fishes may store additional fat to provide extra energy. Migrations involve expending a lot of ennot directly behind another fish but halfway in between the two fish in front of it, and separated from its neighbor by one-half to one body length. This allows each individual to gain a boost from the water currents produced by the tail movements of the fish in front. Imagine pushing a paddle through water: on either side of your paddle there is an eddy, with forward-moving water on the farther edge of each eddy. A fish swimming between and somewhat behind two other fish saves energy because that forward-moving water from each neighbor's eddy gives it a boost. Of course, those on the outside of the school have only one neighbor so have less advantage, and the leaders have no advantage at all. But sooner or later the school will change direction, and they all get a break.

Schooling also provides safety against predators. Fish separated from their school or straggling behind are more likely to be picked off. The school—a large number of silvery, flashing fish—can be confusing to visually oriented predators. Predator satiation also is part of safety in numbers. Being in a school does not guarantee survival, but the odds are much better than being alone. Many predators have strategies to break up schools and pick off individuals. Some predators themselves school, and when a school of predators attacks, a school of prey cannot defend itself.

Schooling can aid in reproduction. At the proper time, it isn't necessary to look for a mate; all they have to do is release sperm and eggs together and fertilization is accomplished. When a school of herring—perhaps a million or more—come into a bay to spawn, the entire bay will become milky from all the sperm.

Question 6: How do fishes communicate?

Answer: Communication is very important for the survival and well-being of animals, and fish are no exception. Fish communicate with each other using visual, auditory, and chemical signals. Optical signals can range from color changes to flashing of a bioluminescent structure. Signals involving color change can be accompanied by movements—postures, approaches, flights—which may show off brightly colored parts such as fins. These parts may be vividly colored only during the breeding season or may be displayed only in the breeding season. Fish also communicate mood by turning pale in fright and getting darker when angry. Some have particular patterns of stripes and spots that can signal certain "emotional" states. Sound can communicate fear, threats, and signals to prospective mates. It can warn intruders in one's territory of the possibility of an encounter. Other sounds are associated with courtship and mating. Fish that produce electricity can also use it to communicate information (see chapter 3, question 16: How and why do some fishes produce electricity?).

Chemical signals are important for communication (see chapter 3, question 9: Do fishes smell?). Cyprinids (minnows) release an alarm substance into the water. If an injured minnow is placed in a school, the whole school flees in alarm. Pouring water from an injured minnow's tank into a school also causes alarm, so it is clear that the warning is a pheromone in the water. John Todd of the Woods Hole Oceanographic Institution examined chemical communication in bullhead catfish, *Ictalurus,* which establish a territory that they defend against intruders. Individual bullheads appear to produce a unique odor so they can recognize each other. When two fish were in an aquarium together, each established a territory and stayed there. If one was removed and a new fish introduced, the resident fish attacked the newcomer. When the stranger was removed and the original fish replaced, the resident again respected the border between the territories, indicating that it recognized the odor of the particular individual with whom it had co-existed peacefully. Chemical communication in this species is also used to distinguish social status, and when the "top fish" lost a fight it produced a different odor and others could sense that it was no longer dominant. These behaviors were the same for blinded fish, so the sense of smell is responsible. This sense enabled fish to form peaceable groups with territories respected by others; when fish were deprived of their sense of smell, they continually fought and didn't adjust. Since this pioneering work, studies of chemical communication have been extended to other species. In many species, females release chemicals into the water before they are ready to lay eggs. The males detect the scent and follow it back to the female to mate. In other species, males release a scent that attracts females for courtship.

The detection of chemical messages from the water can easily be impaired by pollutants that damage the sense of smell. While the whole nervous system is sensitive to pollutants, the olfactory system is especially sensitive since its sensory cells are in direct contact with the water. Since the sense of smell is so important, interfering with it can affect many aspects of fishes' lives. Heidi Fisher and colleagues at Boston University found that female platyfish, *Xiphophorus birchmanni*, normally have a strong preference for chemical cues from males of their own species over a related species, but failed to show a preference when tested in stream water that had sewage effluent and agricultural runoff. R. Mirza and colleagues from McMaster University discovered that in contaminated lakes, yellow perch could detect alarm cues as well as fish from clean lakes, but did not show the normal antipredator behavior that should follow detection of alarm substances. This would make them more likely to be captured by a predator.

Question 7: Are fishes territorial?

Answer: Some fish establish territories for nesting, feeding, or just resting and defend them against members of the same species or other intruders. The population density of territorial fish is determined largely by the availability of suitable habitat, and individuals that successfully defend good territory are likely to be the most vigorous and most likely to grow and reproduce. For example, as juvenile salmon called *parr* move down from their natal stream to deeper water, they find favored spots with abundant food and refuge. They defend these territories via aggressive displays or actual combat and form hierarchies in which fish lower on the pecking order have inferior territories and less chance to survive and complete their migration.

Male mudskippers (see color plate A) indulge in territorial battles where they first display colorful fins. If neither fish is willing to back down, they line up face-to-face and open their mouths as wide as possible, then push against each other until one or the other turns and looks away, signaling defeat. Though this looks like a kiss, it is far from affectionate! Male Siamese fighting fish, *Betta splendens* (see color plate F), are unusual because in the confined conditions of a fish tank, they will inflict considerable damage on each other; in nature, the loser can flee before it is too late. Fighting in most species is usually ritualized displays; physical harm is rare.

Damselfish guard areas on a coral reef rich with algae, their preferred food. I have several times encountered territorial damselfish, a few inches long, aggressively defending their patch and trying to chase away a snorkeling intruder (me) who had no interest in eating their algae. The territories defended by the damselfish *Eupomacentrus planifrons* were studied in detail by Susan Brawley of the University of California, Berkeley. She found that when damselfish were removed from an area, the amount of some algae declined, suggesting that damselfish territoriality may influence overall reef productivity.

Masanori Kohda of Osaka City University studied territoriality in several species of male cichlids in Lake Tangankiya (Africa) and found that they maintained different-sized territories for different purposes—swimming, nesting or feeding. Males swam, foraged, and courted females in restricted ranges containing nest sites that were separated from those of other males. They attacked competitors near the border of the mating territory, while they attacked potential egg eaters only around the nest site (an area within the mating territory) and attacked food competitors in the feeding territory, an area inside the mating territory but away from the nest.

The red grouper stays in a very restricted area that it takes meticulous care of. It excavates and maintains complex, three-dimensional structures that provide habitats for the spiny lobster and many other commercially important species in the Gulf of Mexico. Felicia Coleman of the University of Florida and colleagues watched a red grouper work to remove sand from the sea floor, exposing rocks important to corals and sponges and the animals they shelter. These fish reshape the flat bottom into a three-dimensional structure, which is enhanced by the settlement and growth of corals and sponges. The fish dig holes and maintain them by carrying mouthfuls of sediment from the center of the pit to the periphery and expelling them and then brushing off the rocks with their tail fins. The sites serve to attract mates and prey for the grouper as well as species such as spiny lobster and cleaner shrimp (see this chapter, question 13: How do fish interact with invertebrates?). Groupers are sedentary and are important "ecosystem engineers" of their habitat.

Question 8: Are some fishes naturally aggressive?

Answer: Some fish are naturally aggressive and form social hierarchies or "pecking orders," where some become dominant over others, with the order of dominance related to aggressiveness and size. The largest fish tends to be at the top and can chase the others, the second largest can chase everyone except the largest, and so forth. Aggression may consist of displays like erecting fins or other specific postures rather than actual combat, though fighting can sometimes occur to establish dominance. Submission is shown by lowering fins and assuming a head-down posture. These relationships in which all fish know their place have the advantage of reducing fighting.

In a community tank, one fish may constantly pick on another one of either the same or a different species. The more crowded the fish are, the more likely it is that aggression will develop. Creating a number of hiding places in the tank can help to establish a balance so that aggressive fish will leave most of the others alone. When a new aggressive fish is added to an aquarium with an already-established, territorial fish, the established fish will probably fight to protect its territory (the whole tank). Placing a mirror outside of the aquarium so the aggressive fish can see itself may occupy it with its own reflection and give the new fish time to establish itself without too much hostility. Some aggressiveness (among cichlids, for example) is food related, and aquarists can reduce aggression by feeding the fish less but more often, so they are less likely to feel threatened and competitive. Stocking a home aquarium with fish that have different niches (different behavioral characteristics, foods, habitat preferences) and giving them lots of room and plenty of territories and hiding places will help to reduce aggression. Some aquarium fish, however, are always aggressive toward other species and simply don't belong in a community tank; they will be fine if kept in tanks of their own.

A recent study showed that certain coral reef fishes undergo behavioral/personality changes in warmer water. Damselfishes became more aggressive, increasing their levels of activity, boldness, and aggressiveness, when the temperature was increased by only a few degrees.

Question 9: Do fish sleep?

Answer: Since fishes lack eyelids, they cannot close their eyes (a characteristic of sleep in many animals), but almost all fish sleep. They spend part of every day or night in an energy-saving state that can be categorized as rest, if not precisely as sleep. For any animal, a sleeplike state includes a partial or complete loss of consciousness and a slowing of bodily functions—the heart rate and breathing slows, and muscles relax. Sleeping fish are relatively immobile and are less sensitive to external stimuli. At night, herring and tuna rest motionlessly in the water, and bass and perch rest on or under logs. Rockfish and groupers rest against rocks, bracing themselves with their fins. Reef fishes find a safe spot in the coral to hide and rest. Some actually build a nest in which to sleep. Some parrotfish hole up in a crevice and envelope themselves in a cocoon of mucus. They slumber so deeply that a diver can hold them while they sleep. Many species of shark "sleep" while swimming slowly, because they must keep moving in order to breathe. Minnows, active in schools during the day, scatter and remain motionless in shallows at night. Squirrelfish on coral reefs "rest" or "sleep" during the day under ledges and are active at night. Nocturnal freshwater catfish swim up under a log or riverbank for shelter during the day.

Emmanuel Mignot and Tohei Yokogawa of Stanford University studied sleep patterns of zebrafish, *Danio rerio,* in aquaria; they observed that when the fish appeared to be asleep their tails drooped and they rested either just beneath the surface or at the bottom of their tank. They studied the contrast between the sleep patterns of normal zebrafish and mutant zebrafish that lacked functional hypocretin receptors, which seem to be involved in sleep regulation in mammals, and found that the mutants got less sleep than the normal fish. Scientists are studying other animals in order to understand the relationship between sleep and hypocretin receptors, as malfunction in this system causes the sleep disorder narcolepsy in mice.

Question 10: Do fishes hibernate?

Answer: Hibernation is the process of spending part of the winter in a relatively dormant state, apparently as protection from cold when normal body temperature can't be maintained and food is scarce. Since fish are cold-blooded, their body temperature is the same as their environment. When it is warm, they are active, feeding, and growing, but as the water temperature drops, fish slow down. Many become dormant in winter because cold temperature simply lowers their metabolism. But that is not hibernation. However, one fish has been found to actually hibernate. Antarctic cod, *Notothenia coriiceps,* "take a nap" during the long winter, according to Hamish Campbell of the University of Queensland, Australia. The cod hunker down on the seafloor, reduce feeding, and slow their heart rates, probably as a way to survive the long dark winters when finding prey is difficult. This is the first time fish have been seen to become torpid—a state similar to hibernation—as part of an annual cycle. They become twenty times less active in winter than summer and undergo this metabolic slowdown regardless of water temperature. Hibernation is triggered by seasonal changes in light levels, which vary far more than water temperature in the always-frigid Antarctic. The fish use the short Antarctic summers to gain sufficient energy from feeding to tide them over in winter when they hibernate and reduce energy requirements to the bare minimum. Every week or so they wake up and swim around for a few hours, which is similar to "denning" in bears, where the hibernation isn't so deep and they spend some time awake.

Other fish *aestivate* in hot weather, which is also a slowing down of the body's functions. South American and African lungfish are capable of an amazing aestivation that allows them to live without water for up to three years. When a pond dries up during the dry season, the fish burrows into the mud, then secretes mucus over its entire body. The mucus dries into a sack that holds moisture in. Even when the mud dries completely, the lungfish stays moist and breathes air through a hole in the mud (see chapter 2, question 4: Can any fishes live out of water and breathe air?).

Question 11: Do fishes learn?

Answer: The modification of behavior that we call learning is seen clearly in fishes. It is involved in the return of migrating fish to their home ranges. Recognition of mates, prey, and predators requires some modification of behavior. When pecking orders are established, it is necessary to remember who is above you and who is below you (who you can pick on). General observations indicate that fish learn quickly and have reasonably good memories. Fish in home aquariums learn to recognize the fish food and cluster around before it hits the water. We have brought wild mummichogs into the laboratory; initially, when a hand appears above them to drop in food, they panic and hide on the bottom as if it were a predatory bird. After several feeding sessions, they overcome this innate response and come to the surface instead. The ability of fishes to learn to eat new kinds of food is important for aquaculture. If they can learn to eat an artificially formulated diet, it is easier and cheaper than relying on natural foods alone. The angelfish, *Pterophyllum scalare*, learned to come to different parts of a bare tank at different times of day in order to be fed, in an experiment by L. Gomez-Laplaza and E. Morgan of the University of Oviedo, Spain. For three weeks, food was delivered in one corner of the tank in the morning and in the opposite corner in the afternoon. Without any stimulus such as the sun or tides, they were able to distinguish between feeding sites and go to the right place at the right time, showing that not only do they learn but they can "tell time."

Haddock, *Melanogrammus aeglefinus*, currently an endangered species of the North Atlantic, can learn how to avoid getting caught in nets. While scientists remain unsure exactly how the learning actually takes place, they believe that experience of a net when the fish are young enough to pass through it helps them learn how to avoid it later. Chris Glass and colleagues at the Manomet Center of Conservation Sciences in Massachusetts trained haddock to swim across a pool toward food on the other side. Later, a net with large mesh was placed across the pool, initially deterring the fish—until a few swam through it to the food. Each time the experiment was repeated with the mesh getting smaller but still big enough to let the fish through, the experienced fish led the way and the rest would follow, apparently learning from watching more experienced fish. At sea, young fish are thought to slip through the mesh to escape nets. When they are adults they are more likely to avoid the nets altogether.

Question 12: How do fish interact with other species of fishes?

Answer: The most common interaction is between predators and prey, but there are other interactions that are not harmful to either species. Remoras have a suction disc on the top of their head that attaches them to larger fishes (fig. 15). This gives them a "free ride" and easy access to food that escapes from

Figure 15. Remora, *Remora* sp., sticking on the wall of a tank. *(Photo by Peddrick Weis.)*

the predator's jaws. The larger fish, often sharks, are neither harmed nor helped by the "hitchhiker."

One of the best-known associations between fishes is cleaning symbiosis, which benefits both species. Cleaner fishes (usually reef fish called wrasses) go over the bodies of larger fishes, picking off external parasites, which they eat (see color plate E). The large fishes, including moray eels, snappers, groupers, and barracuda, which could easily eat them if they chose to, let the cleaners pick all over their bodies, including inside their gill chambers and mouth, which they open, allowing the cleaner to remove parasites and any dead tissue. This improves the health and well-being of the cleaned species and provides a meal for the cleaners. Cleaners, which are brightly and distinctively colored, establish locations (termed *cleaning stations*) where "customers" congregate and await their turn. Customers pose, while spreading their fins and hovering in the water. In some cases, non-cleaning blennies mimic the coloration and behavior of the cleaners to get access to the larger fish, and instead of removing parasites, take a bite out of them. While cleaning symbiosis has been studied mostly in tropical reefs, this may be due largely to the presence of more scuba divers and snorkelers in these areas; cleaning symbiosis has been found in freshwater species, too, as well as in other marine habitats.

"Doctor Fish" in Spas

If you want to know what it feels like to be a fish being cleaned, you can visit a spa that has Doctor Fish, *Garra rufa*. The fish are kept in outdoor pools where they feed on the skin of patients with psoriasis or other skin conditions. The fish consume only the affected and dead areas of the skin, leaving the healthy skin alone. You immerse your feet, hands, or, if you are brave enough, your entire body in a warm pool with hundreds of hungry minnow-sized fish that zoom in on crusty, flaky, or dead skin and chomp away at it, revealing the fresh layer beneath. Doctor Fish are attracted to dead skin, calluses, and corns. According to the spas and their enthusiasts, you emerge refreshed, healthy, and glowing. These fish, which are native to rivers of the Middle East, are now found in spas in many countries. In 2008, the first widely known Doctor Fish pedicure service was opened in the United States in Alexandria, Virginia.

Question 13: How do fishes interact with invertebrates?

Answer: The most common interaction again is eating—mainly fishes eating invertebrates, but some invertebrates (e.g., squids, large crustaceans) eating fish. There are also many invertebrates that are parasites of fishes (discussed in chapter 6, question 2: What kinds of parasites do fishes have?). Other relationships are more cooperative. In addition to cleaner fish, there are also cleaner shrimp that perform the same function on coral reefs (see color plate H). In tropical regions ghost shrimp (which have poor vision) dig burrows that they share with certain gobies, which stand guard.

One of the best-known relationships between fish and an invertebrate is that of clownfish or anemonefish, genus *Amphiprion*, with sea anemones (see color plate E). Anemones, stationary relatives of jellyfish, have a stalk, attach to a hard surface, and have stinging tentacles that can paralyze smaller organisms that touch them. The tentacles contain fluid-filled cells (*nematocysts*), each with a microscopic harpoon-like venomous structure, similar to that of jellyfish. The anemone uses the nematocysts to defend itself against predators, and its venom is strong enough to kill small fish, but clownfish live among the tentacles without being harmed. Particular species of clownfish generally live with particular species of anemones. Clownfish eat algae, copepods, isopods, and zooplankton as well as uneaten prey captured by the anemone, and the wastes from the clownfish, in turn, provide nutrients to the sea anemone. Recent research has indicated that ammonia excreted by the fish helps the anemone by increasing the abundance of symbiotic algae, *zooxanthellae*, that photosynthesize inside an anemone. Modi Roopin and Nanette Chadwick of Auburn University found that anemones grown without either ammonia or anemonefish had fewer zooxanthellae and smaller body size. The fish also feeds its host, grooms it to keep it free of parasites and debris, and protects it. It has been suggested that the movement of the clownfish increases water circulation around the anemone, sending it more plankton to eat. In addition to providing food for the clownfish, the anemone also provides safety with its tentacles. Potential predators know not to go near the anemone's tentacles. This kind of relationship in which both partners benefit is called *mutualism*.

But how do the clownfish avoid being stung by their host? The mucus that they secrete seems to be involved. Since certain species of clownfish have coevolved with certain species of anemones, they may have acquired a resistance to the nematocysts of their host anemone. Experimentation has shown that *Amphiprion percula* may develop resistance to the toxin from *Heteractis magnifica*, but it is not totally protected; when its skin, devoid of mucus, was exposed to the nematocysts of its host, it died. Acclimation must occur before an individual fish is resistant to the tentacles of its host. Acclimation is accomplished by the fish rubbing its ventral side and fins on the tentacles. Several other fish species live among tentacles of various jellyfish and of the Portuguese Man o' War. However, *Nomeus*, the fish that lives with the Man o' War is vulnerable to the sting and survives only by avoiding the tentacles.

Question 14: How do fishes interact with plants?

Answer: In addition to eating them, fishes use plants as habitat and protection. Some live among roots of mangrove trees or in adjoining sea grass beds. Some estuarine fishes go up onto the surface of intertidal salt marshes when the tide comes in. The mummichog or killifish (fig. 16) does most of its feeding on small organisms up on the marsh during high tide.

Aquatic plants can increase fish abundance, particularly in areas that did not previously have many fishes. J. C. Borawa and colleagues from the North Carolina Wildlife Resources Commission found that fish densities increased greatly after the invasive plant *Myriophyllum spicatum* became established. K. J. Killgore and colleagues from the Army Corps of Engineers Waterways Experiment Station estimated that fish density in the Potomac River was two to seven times higher in areas with aquatic plants than areas without plants. In addition, the presence of fishes can help plant growth, even terrestrial plants. Tiffany Knight of Washington University found a correlation between the presence of fish in ponds and well-pollinated Saint-John's-wort

Figure 16. Mummichog (or common killifish), *Fundulus heteroclitus*. (*Courtesy of NOAA.*)

nearby. She studied ponds with fish and without fish and found that plants near the fish ponds were better pollinated. The reason? Fish eat dragonfly larvae in the ponds and reduce dragonfly populations. Adult dragonflies eat bees, moths, and flies that pollinate flowering plants. So more fish leads to fewer dragonflies, which leads to more pollinators, which leads to more plants.

Some freshwater fishes deposit eggs in plants. Egg-scatterers (e.g., minnows) swim into clumps of plants such as *Elodea* or *Vallisneria* and scatter their eggs. The plants hide the eggs from predators and prevent them from drifting away. Fish with sticky eggs can anchor them onto leaves so they will not fall to the bottom. Other plants are modified when fishes remove plant material (including live, rooted plants) from the spawning site in order to produce a more sterile environment. Sticklebacks select aquatic vegetation and form it into a nest for spawning and early rearing of fry. Mummichogs lay eggs at the base of stems of intertidal salt marsh grasses, so the embryos develop on the marsh surface mostly in the air, but where they can stay moist.

A number of fishes that live among seaweeds or sea grasses are camouflaged to resemble the plant. Shrimpfish, family Centricidae, orient head-down among blades of sea grass, vertically oriented coral, or sea urchin spines. The fish has a black stripe down its transparent body that helps it blend in. Most striking among seaweed residents are the Sargassum fish, *Histrio histrio*, which live among rafts of floating Sargassum weed in the Sargasso Sea, and the leafy sea dragon (see color plate F). A seahorse relative, the leafy sea dragon, *Phycodurus eques*, from southern and western Australia, has long leaf-like protrusions coming from all over the body that serve as camouflage and allow them to blend in with the seaweed. A dragon swims using small transparent fins that undulate to move the fish slowly through the water, looking like a piece of floating seaweed. Tiny, translucent pectoral fins along the sides of the head are used to steer, and dorsal fins, one at the junction of the neck and body and one at the base of the tail, provide propulsion. Sea dragons swim horizontally, unlike seahorses. They have a long snout with a small terminal mouth that sucks in plankton and other small organisms. In the past decade, most public aquariums have acquired dragons for people to watch. Seahorses themselves are frequently associated with sea grasses, flowering plants in shallow coastal areas. As sea grasses are decreasing throughout the world due to water-quality problems and coastal development, there is concern about the future of species that depend on them.

FIVE

Fish Reproduction

Question 1: How do fishes mate?

Answer: One of the most striking things about fish reproduction is the sheer number of offspring fish are capable of producing. During its reproductive life, a single female of some species can produce millions of eggs. If all of them were to survive, the world would be overrun with fishes, which is clearly not the case. Fish reproduction represents a kind of constant battle between the forces of reproduction and predation and the fish's environment.

Given the large number of fish species, it is not surprising that they have evolved very diverse reproductive methods that allow them to be successful under a variety of conditions. One of the most common strategies among egg-laying, or *oviparous,* fish is the fertilization of large numbers of eggs at a time through *spawning.* Environmental cues trigger the simultaneous release of the male's sperm and the female's eggs into the water in a milky cloud where fertilization occurs. Some fishes pair off to spawn, while others engage in mass spawning, in which all of the fish of a species in an area come together to release sperm and eggs. Although the spawning process sounds a bit "hit or miss," it works because fishes that spawn like this often produce millions of eggs each, even though only a tiny percentage survive. Species that use this method include many food fishes such as tuna, cod, mackerel, pollack, hake, halibut, and herring.

After spawning, most fishes leave the eggs behind and let the embryos fend for themselves. Some species, however, place the eggs somewhere where they will be protected, to increase their chances of survival. While floating eggs move with the current, eggs that are heavier than water (*demersal*) may be covered with a sticky substance and placed on a rock, inside a shell, on vegetation, in a constructed nest, or even on one of the parents. Some fishes have more complicated ways to fertilize their eggs. Among cichlids like tilapia, for example, a male entices a female into his territory, where she lays about two hundred eggs that she picks up and holds in her mouth. The male has spots on his anal fin that resemble her eggs, and as she tugs on his anal fin in an attempt to pick up those last few "eggs," he releases his sperm into her open mouth, where they are fertilized and develop for a week or two until they hatch. In seahorses and pipefish, mating involves the female inserting her ovipositor into the male's brood pouch to deposit her eggs, where he fertilizes them and then incubates the embryos. Among the egg layers, females are often larger than the males to accommodate so many eggs, and in deep sea anglerfish this size difference is extreme. As a juvenile, the male finds a female and attaches to her underside, where he remains as a small parasite (see fig. 5, in chapter 2), providing sperm in exchange for nutrients. In the deep sea, potential mates are few and far between, so these fish are joined for life.

Instead of egg-laying, several types of fishes copulate, with fertilization and embryo development taking place inside the female. Internal fertilization requires modification of behavior and anatomy. Sharks and rays copulate using modified pelvic fins, or claspers, which transfer sperm into the female. In Poeciliids (guppies, platyfish, swordtails) (see color plate F) the anal fin is an elongated into a *gonopodium* for transferring sperm. Brian Langerhans of Washington University studied mosquitofish, Poeciliids, and found that females spent more time with males that had a large gonopodium and were more likely to mate with them. However, when predators were present, these males were more likely to be captured because they couldn't escape as fast. It seems that bigger is better for mating, but smaller is better for surviving. Among swordtails, *Xiphophorus helleri,* that have an elongated lobe of the tail fin (see color plate F), females prefer to mate with males with longer swords, so sword length is important in mate choice. Poeciliid females store sperm from one mating to use for several broods. Shark and poeciliid embryos develop to an advanced stage before being born. The embryos may be nourished by yolk in the egg (*ovoviviparous*) or may get nutrition from the female (*viviparous*) through a structure that serves the same function as a placenta in mammals.

Species with internal fertilization and incubation face a trade-off. Retaining the embryos inside the body of the adult protects them and the adults don't have to migrate to specific breeding sites or watch or protect the eggs. Copulation generally assures fertilization of the eggs, so fewer eggs are wasted, but this also limits the number of eggs that can be produced and fertilized.

Question 2: How does a fish attract a mate?

Answer: Mating behaviors, the location, timing, and signals of courtship are extremely varied. Fishes that migrate gather at a spawning area when they are ready to breed. They signal their reproductive readiness by changing color during the breeding season. This "nuptial coloration" is controlled by hormones. In some species, males wait at a site and court any female in the area. In others, the male courts and mates with only one or a few females. Olfaction, sound production, and bioluminescence can all play roles in courtship. Male dwarf stonebashers "sing" courtship songs at night, songs that are specific to their own species to attract females. Mormyrid couples use specific electrical signals to court and engage in an electric duet.

Typically, the male has a stylized approach that displays his form, coloration, or posture. The male's appearance and behavior as well as his territory are important in female mate choice. If she is receptive, she approaches or allows him to approach her. There may be nuzzling, head-butting, nips, side-by-side swimming, "dancing," or other interactions before mating. Some

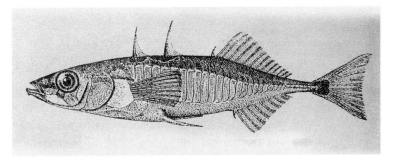

Figure 17. Threespine stickleback, *Gasterosteus aculeatus*. *(Courtesy of NOAA.)*

cichlid pairs grab each other's mouths and wrestle to test their compatibility. If there is a nest, the male must persuade her to go in and deposit her eggs.

The ritualized courtship and mating behaviors of the threespine stickleback, *Gasterosteus aculeatus* (fig. 17), were the focus of classic research by the Dutch scientist Nikolaas Tinbergen, who won a Nobel Prize for studies of animal behavior. Each male stakes out a territory in shallow fresh water that he defends against intruders, especially other males. Then he digs a pit in the sandy bottom by taking sand away in his mouth, builds a nest from algae and plant material that he collects and coats with a sticky secretion from his kidneys, and shapes it into a mound. He then wriggles through the mound creating a tunnel. This is the nest. He then develops a bright red coloration and begins to court females, who are full of eggs and have swollen bellies. When one approaches, he swims toward her in a zigzag motion. If she responds, he goes toward the nest and she follows. He pokes his snout into the entrance, then turns on his side and raises his spines, and she enters the nest to lay her eggs while he prods her near her tail. After the eggs are laid, she leaves (or is chased away). He fertilizes the eggs, guards the nest, attacks all intruders, and fans water through to improve the oxygen supply. The time spent fanning increases as the eggs develop. After the eggs hatch, he keeps the brood together for a few days.

Through a series of elegant experiments, Tinbergen found that each phase in the cycle was triggered by a few visual

Your Cheatin' Heart

Deceit, a regrettable part of human dating and mating, also occurs in fishes. Some male longear sunfish (see color plate D), known as "sneakers," are generally younger and less colorful than nest-building males. The sneaker male mimics the appearance of a female and will hide near an active nest and dash into the nest and release his sperm while the dominant male is spawning with a female. Atlantic molly (*Poecilia mexicana*) males prefer to mate with larger females, but when other males are present, they may disguise their choice by feigning interest in a different female. This appears to be an attempt to divert rivals' attention away from the female of choice and reduce competition with other males. Suzannne Gray of Queen's University, Ontario, found that males of the sailfin silversides *Telmatherina sarasinorum* followed courting pairs of a related species, *T. antoniae,* chased off the males, courted the females, and enticed them to spawn with them—and then ate the eggs!

characteristics ("signs") from the other fish rather than by the male fish itself. For example, he found that the male's red color was the stimulus for attacking, and males would attack red items that did not look at all like a fish. He was clued into the importance of the red color after seeing males becoming agitated whenever a bright red vehicle passed by. It was then that he (Tinbergen, not the fish) started analyzing males' responses to red versus non-red fish models, and even to a red blob on the end of a rod. He likewise found that it was the rounded belly of the female that triggered the courtship behavior; males would court other males that were swollen with food. Likewise, females follow the red color, even if it isn't a male stickleback, and will release eggs when prodded with a glass rod.

Seahorses (see color plate D) undergo a complex courtship in which they may change color, swim side by side holding tails or grip the same strand of sea grass with their tails and wheel around. They eventually engage in their eight-hour courtship dance, during which the male pumps water through the egg pouch on his trunk, which expands and opens. When the female's eggs reach maturity, she and her mate drift upward facing each other, often spiraling as they rise in preparation for actual mating, when she deposits eggs into his pouch.

Question 3: Do fishes build nests?

Answer: While the young in most marine fish are left to fend for themselves, a number of freshwater fishes build "nests" in which the young are incubated and protected. The variety of nests and nest-building techniques are intriguing. Some species simply clean off the surface of a rock for depositing eggs. In streams, darters stick their eggs on the bottom of rocks, where they adhere and are protected and well-oxygenated. Salmon and trout deposit eggs in gravel depressions (*redds*) that they dig in stream bottoms and then cover up. In ponds, fathead minnows spawn on the underside of old logs. The males develop a pad containing mucus-secreting cells on the top of their heads that they use to rub algae and debris off the wood before the eggs are deposited. Killifishes in temporary ponds in Africa and South America bury their eggs in the mud, where they stay in a dormant state when the pond dries up. When the rainy season comes, the embryos hatch, grow up, and lay their own eggs in the mud in four to eight months. Such fishes are known as annuals.

While trout and salmon abandon their nests or die shortly thereafter, other fishes stay around to tend their nest. Nest guarders protect their eggs and young (*fry*) and typically produce fewer eggs because survival is higher. It is common for males to care for the developing eggs and young. They keep the eggs free of debris by fanning them, keeping a current of oxygen-rich water passing over them. They are territorial and drive others away from the nest.

Gouramis and Siamese fighting fish build "bubble nests" (see color plate F) in which each egg is encased in a bubble of air blown from mucus by the father. Bubble nesters continue parental care after they construct the bubble nest and spawn. The eggs float up into the bubble nest, or are carried there in the male's mouth. He then protects the brood by chasing away the female and any other intruders. He will retrieve any eggs or fry that fall from the nest and keep it in repair until they hatch in twenty-four to forty-eight hours. For the next few weeks, the fry will stay nearby, being tended by their father.

Other species build more recognizable nests by altering the substrate or using aquatic vegetation to provide a place for the eggs. Sunfishes dig depressions in the bottom and tend their eggs. Their nests attract redfin shiners who also deposit their eggs in the sunfish nest to receive the benefits of the parental care of the sunfish. Other fishes, like the stickleback, use aquatic vegetation for their nests (see this chapter, question 2: How does a fish attract a mate?). An unusual marine fish that makes and guards a nest is the Antarctic plunderfish; both parents guard the nest for up to five months, development being very slow at such cold temperatures.

Some species carry the embryos around after spawning, making a nest out of a part of their own bodies. The most common of these are mouth brooders, mostly cichlids like tilapia, which carry the eggs in their mouths during embryonic development, refraining from eating anything during that time. Unfortunately, this type of nest does not completely protect the eggs. Some predatory species steal the eggs from their safe haven by ramming into the mother or sucking the eggs from her mouth. In cichlids, the young continue to use their mother's mouth as a refuge for some time after hatching. Mouth brooding is also seen in the tropical marine jawfishes, *Opisthognathus,* which, with their googly eyes, appear to have been the inspiration for the Muppets. They live in burrows from which they pop up as Oscar the Grouch does from his garbage can. They carry sand or pebbles from one location to another, or remove them from the burrow. After mating in the burrow, the male takes the eggs into his mouth, putting them down only when he comes out to eat. Eggs can be seen in the half-open mouths of brooding males (fig. 18), where currents of water move them around. Seahorses

Figure 18. Jaw-
fish, *Opisthogna-
thus*, incubating
eggs in his
mouth. *(Photo
by Phil Sokol.)*

and pipefish incubate their embryos in the father's pouch, and
clingfish, *Platystachus cotylephorus*, lacking a pouch, brood their
eggs by sticking them to their bellies. Some fish make their nest
in another organism. The European bitterling, *Rhodeus amorus*
(a minnow), deposits its eggs via a long tube (ovipositor) into
the gills of freshwater mussels, where they are protected from
predators until they hatch. Bitterlings are territorial only during the breeding season when they are guarding a nest.

Question 4: How many offspring are born to a fish each year?

Answer: Fish with mass spawning broadcast millions of eggs
into the water and are by far the most prolific. However, only
a minuscule fraction of those embryos will survive past their

larval stages. The record-holding species for egg production is the greasy gopher, *Epinephelus tauvina*, a large grouper of the Indo-Pacific that aggregates to spawn and whose females each may produce as many as 340 million eggs each season. Black marlins produce between 67 million and 226 million eggs each. Numbers of only one or two million eggs per season are more common. The egg size relative to the size of the female is important—many more tiny eggs can be produced than large eggs. However, larger eggs develop into larger larvae, which will have a better chance of surviving.

The lifetime reproductive success of an individual depends on the number of eggs it produces that survive to reproduce themselves. In assessing this we need to know not only how many eggs a female produces in a reproductive season, but how many reproductive seasons she lives through. Most, if not all, of the eggs any individual female produces will be eaten or destroyed as eggs by bacterial and fungal diseases. If every pair of breeding fish leaves behind just two successfully breeding adults the population will remain stable.

Size determines the number of eggs a female can produce. Older, larger females produce more eggs than younger ones. A female Atlantic cod, *Gadus morhua* (fig. 19), weighing 5 kilograms will produce about 2,500,000 eggs, while one weighing 34 kilograms will produce about 9,000,000. It has been found that the "big old fat females" not only produce more eggs, but eggs of higher quality that are more likely to develop into larger, healthier larvae. For this reason, there is worry that fisheries that go after the largest fish are removing the best breeders from the population.

Sharks and rays with internal fertilization and development have a smaller number of offspring. The great white shark produces about seven, the black-spotted shark, one to two; the bristly catshark, about two; and the Caribbean sharpnose shark, two to six young. Reducing the numbers further in some species is the sharklike behavior that leads larger embryos to devour their smaller "womb mates." The low reproductive rate of sharks

Figure 19. Atlantic cod, *Gadus morhua*. *(Courtesy of NOAA.)*

makes them particularly vulnerable to overfishing because their populations cannot readily bounce back.

In between the sharks and rays with their few live-born babies and those that broadcast vast numbers of tiny eggs are those that lay relatively few—dozens to a few hundred—eggs. To compensate for the relatively few eggs, the females hide them from egg-eaters (salmon, for example, lay them in gravel beds). These eggs are individually much larger than the millions of broadcast eggs, so they will hatch into larger larvae that are more likely to avoid predation. The best chance for survival of small batches of eggs is to lay eggs in nests and guard them.

Question 5: How does a fish embryo develop?

Answer: Before development can start, the egg must be fertilized. While eggs of most animals can be penetrated by sperm anywhere on their surface, eggs of bony fishes are unusual in that their outer membrane, the *chorion*, has a single spot, a small opening called the *micropyle*, through which sperm can enter.

After fertilization, the egg reorganizes so that there is a single cell (zygote) perched on top of a mass of yolk (fig. 20a) that provides nutrients. This cell undergoes numerous divisions (*cleavage*) until there is a pile of cells at this location (fig. 20b). Then the cells start spreading (*epiboly*) over the yolk (fig. 20c). After about half the yolk surface is covered, a thickening occurs along the margin, and the long axis of the body gradually becomes visible, curled around the yolk (fig. 20d). Once the body axis is laid out, segmentation, followed by the gradual appearance of eyes and other structures, takes place (fig. 20 e and f). Fish eggs are transparent, so it is marvelous to watch them develop. The time from fertilization to hatching may be as little as a day or two, or it might take months, depending on the species and the temperature. Those that hatch quickly still have much of their yolk left so that they don't have to feed for several days up to a week. These quick-hatching species may not even have fully developed jaws, so they couldn't feed anyway. Since verbal descriptions of fish development are nowhere near as good as watching it, you can watch time-lapse photography of zebrafish development on You Tube: http://www.youtube .com/watch?v=ahJjLzyioWM&feature=related.

Fishes that develop within the body of the female are not easy to observe and vary considerably in the degree to which the female provides more than just space for them. Guppies, platys, and mollies provide a little nutrition for the embryos. Some sharks and skates retain their embryos for a short time and then release them in a tough, rectangular capsule with extensions on each corner to attach to seaweeds. Skate egg capsules occasionally wash up on ocean beaches; enchanted beachcombers nicknamed them "mermaids' purses." Some sharks are *ovoviviparous* and simply retain their young in the posterior segment of the oviduct without providing nutrients, while others are *viviparous,* and the female provides nutrition like mammals do. The uterus develops secretory structures that produce nutrients. In stingrays, after the yolk is depleted, the uterine lining develops *trophonemata,* long filaments that secrete a nutritive fluid that is absorbed by the embryo. The secretion, called "uterine milk," is

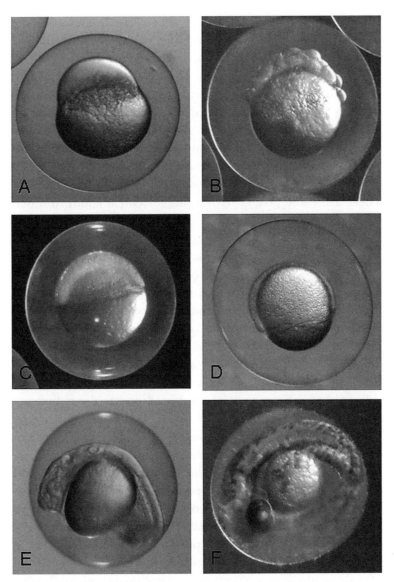

Figure 20. Embryonic development of the zebrafish, *Danio rerio. (Photo by Steve Baskauf.)*

rich in proteins and lipids, similar to milk of mammals. In some sharks the yolk sac develops into a functional placenta, while in others, after the yolk is used up, the embryos develop teeth and feed on eggs inside the uterus or on their siblings. In the sand tiger shark, only one embryo survives in each uterus (sharks have a left and right uterus) so litter size is never more than two. The one remaining fetus per uterus grows very large. Other species that do this are mako, basking, and thresher sharks. White shark embryos do not cannibalize their siblings but do eat eggs.

Seahorses and pipefish embryos develop in the male's pouch. There are three general types: simple ventral gluing (no real pouch), two pouch flaps, and a completely sealed sac. These provide different degrees of nutrition, aeration, and protection. In the species with the most paternal involvement, the embryos embed in the pouch wall and become enveloped with tissues, similar to the uterus of a mammal. The pouch contains a fluid that bathes the eggs and provides nutrients and oxygen while also removing waste products. At the end of gestation, the male goes into "labor," pumping and thrusting for hours to release his brood. Young are miniature adults, independent from birth. Jennifer Ripley and Christy Foran of West Virginia University showed that the structure of the brood pouch alone does not necessarily tell you how much nutrition the father provides. They found differences in nutrient provision to embryos for two closely related pipefishes with the same brood pouch structure. Northern pipefish females produce nutritionally poor eggs, and embryos implant in the brood pouch adjacent to blood vessels. In contrast, dusky pipefish have only a loose connection between embryos and brood pouch tissues but also have undeveloped eggs in the pouch that serve to provide nutrients to feed the developing embryos. This comparison shows very different degrees of male involvement even when the pouch types are similar.

Question 6: How do fishes raise their offspring?

Answer: For the vast majority of fishes, the answer is—they don't. Even live bearers tend to view small fish, including their own babies, as dinner. However, some nest builders, mouth brooders, and pouch bearers guard the eggs and then the larvae for some time after hatching. In most cases where the young are cared for, among catfishes, sunfishes, cichlids, and others, for example, the males take the lead. The male betta (Siamese fighting fish) (see color plate F), after guarding the bubble nest, watches his young and signals danger by shaking his pectoral fins near the surface of the water. The waves are perceived by the young, who swim toward him in response. He can then scoop them up in his mouth and carry them back to the safety of the bubble nest.

Both sexes of cichlids provide extensive parental care. Convict cichlids and jewel cichlids dig a pit in the sand and then both parents gather up their young at night, catching two or three at a time in their mouths and spit them out into the pit. They do this repeatedly until all fry are "put to bed." This happens indoors in constant light, so apparently the fish have an inner sense of time and are not just responding to environmental changes. As reported by Konrad Lorenz in his book *King Solomon's Ring*, when this bedtime activity was taking place, he dropped some food (a piece of earthworm) into the tank. The male grabbed it, then he saw one of his fry swimming by and took it into his mouth too. And then the fish stopped—what to do? After a few moments he spat out both the worm and the fry. Then he ate the worm, while keeping an eye on the fry nearby. After he swallowed the worm, he picked up the fry and brought it back to the nest. Problem solved. Perhaps fish are more intelligent than we humans give them credit for.

Although the female tilapia *Oreochromis* produces relatively few eggs, she provides an unusual degree of protection, so a high percentage survives. Fry swim and feed, but stay near the female. When they sense danger, they swim back into her mouth for protection. Communication between fry and parents has been seen in a number of cichlids. Frequently, this communication is based on movements, such as shaking and pelvic fin flicking. Parental cichlids help their fry find food by turning leaves or digging with a fin. Communal parental care, where multiple pairs care for a mixed school of young, has been observed in the midas cichlid and the redbreast tilapia. Fry of the fairy cichlids, which live in large groups, are protected by adults and by older juveniles from previous spawns.

Some cichlids, including discus, *Symphysodon* spp. (see color plate G), feed their young with a skin secretion from mucous glands. Both parents secrete protein-rich mucus that contains many proteins not present in mucus of non-breeding fish. Production of this mucus is stimulated by the hormone prolactin, the very same hormone that stimulates development and secretion of milk in mammals. Larval feeding on parental mucus increases until about fifteen days of free-swimming and then decreases as they begin to eat other food.

Question 7: How do fishes develop as they grow older?

Answer: Newly hatched fishes of oviparous species are often tiny and very poorly developed. Their large yolk sac provides food for them for several days until they can feed on their own. But soon they must capture microscopic zooplankton at a time when their nervous system is so immature that they must wait for prey to accidentally approach and swim into their mouths. Larvae may have special respiratory devices, sometimes fins with many blood vessels, because their gill covers are not open yet, and oil droplets for flotation since they are not able to swim well. Planktonic larvae have a very low survival rate—they are easily swept away or eaten by most everything that is bigger than they are. In contrast, newly born seahorses, sharks, and other species with internal development are born in a more advanced stage and are basically miniature replicas of adults.

Larvae of many species tend to look similar to each other but very different from the adults of the same species. In some, the transformation is gradual, but in others there is a clear and striking metamorphosis. Most striking are flatfish larvae—they start out looking pretty much like any other fish larvae, but then the entire structure of the head changes as one eye (along with its nerve) migrates over the top of the head until both eyes are on the same side. As this happens, the fish moves downward to spend the rest of its life on the bottom, lying on its now eyeless side (see color plate D). Matt Friedman of the University of Chicago recently found new fossils that revealed how the positions of the fishes' eyes gradually changed over millions of years. The most ancient ancestors had asymmetrical skulls, but retained eyes on both sides of the head. Intermediate species show how one eye gradually moved across the head so that both eyes eventually ended up on the same side.

Eels change from flat, leaf-like leptocephalus larvae, which may grow quite large, to transparent "glass eels," to juveniles that resemble the adult (see fig. 4, in chapter 2). The leptocephalus larva looks so little like an eel that it was originally described as a new genus named *Leptocephalus* until scientists realized their mistake. A similar discovery has recently been made by David Johnson from the Smithsonian Institution. He and colleagues found that certain deep-sea fishes that had been assigned to three very different families (seemingly quite unrelated) were, in fact, larvae, males, and females of only one of those families, the whalefishes, Cetostomidae. Transformations involve dramatic changes in the head, changes related to feeding. Larvae have small, upturned mouths and eat planktonic copepods. Females have huge gapes with long jaws, allowing them to capture larger prey. Males cease feeding, lose their stomach and esophagus, and apparently convert the energy from the copepods to a massive liver that supports them throughout adult life. No wonder they seemed like totally different beasts!

Individuals have the genetic potential to reach a maximum size, but whether they actually reach it depends on nutrition and the environment. The growth rate is fastest when the fishes are small and decreases as they grow. Growth is slower in the cold winter than in the warm summer. Unlike mammals, fishes do not cease growing altogether, but just grow progressively less as they get older. The study of growth is interesting scientifically, but is also important in the management of fisheries. Since the age of the fish can be figured out by examining the rings on scales, it is possible to calculate growth. Measurement of specimens of different ages can provide an estimate of the growth rate from year to year. Knowledge about growth rate is important for aquaculture, where rapid growth is desirable for fish to reach a marketable size quickly. Growth rates for aquaculture can be increased by selection (continual breeding of the fastest growers) and, more recently, by genetic engineering, creating transgenic fish with more growth hormone.

Question 8: Can a fish change sex?

Answer: Many fish change their sex, some from female to male (*protogyny*), others from male to female (*protandry*). In his studies of the bluehead wrasse, *Thalassoma bifasciatum,* a common reef fish, Robert Warner of University of California, Santa Barbara, found that large males (the only ones with blue heads) (see color plate G) stake out a spawning area where many females visit to deposit eggs. If the male dies, the largest female will change sex and become a male. She begins acting like a male within a few hours and will produce sperm within ten days, by which time she (now, he) will have a blue head. The bluebanded goby also switches sex in response to changes in its social environment. Removal of the dominant male in a group results in the dominant female changing sex. Her behavior changes and her ovary transforms into a testis. Such sex changes, once thought to be rare, are more common than previously thought. In deep sea fishes that rarely encounter potential mates, reproduction may be possible only if one changes sex.

Anemonefishes or clownfishes, genus *Amphiprion* (see color plate E), are protandrous. They live in groups on anemones in a symbiotic relationship (see chapter 4, question 13: How do fishes interact with invertebrates?) Only the two largest fish in the group will mate, the largest female and the second largest, the male. The rest are immature males. As found by Hans Fricke and Simone Fricke of the Max Planck Institute, when the female dies, the largest male will become female, and the next largest male will become reproductive. Although this is well known among biologists, it apparently was not known to the writers of the movie *Finding Nemo*, in which the "widowed" father fish, Marlin, seeks his offspring. In reality, "Marlin" would have turned into "Marilyn."

Question 9: Can a fish reproduce without mating?

Answer: There are two mechanisms that enable animals to reproduce without a mate. *Parthenogenesis* involves development of an embryo from an egg without need for sperm, producing offspring that are genetically identical to their mothers. David Chapman of the University of Miami reported that a blacktip shark in a Virginia aquarium became pregnant despite not having been around a male for many years. The "virgin birth" was verified by genetic analysis. He had earlier verified another fatherless birth to a bonnethead shark at an Omaha zoo. According to their reports, sharks are the fifth major vertebrate group with documented virgin births. This allows reproduction in situations where males are scarce or the population is small and dispersed. This may be a good thing, since many shark species have declined dramatically around the world, but has the disadvantage of reducing genetic variation.

Parthenogenesis occurs in a common aquarium fish, the Amazon molly *Poecilia formosa,* which consists of only females and was the first unisexual vertebrate discovered. While they do not reproduce sexually, their eggs can only begin developing when triggered by sperm from a male of a related species, like the sailfin molly. The sperm trigger development but do not fertilize the eggs. Scientists think the Amazon molly probably evolved from sailfin mollies about seventy thousand years ago. Amazon mollies are at a disadvantage because male sailfin mollies prefer to mate with females of their own species, rather than giving sperm to the Amazon mollies.

Another way to have babies without a mate is for an individual to make both sperm and eggs—to be both male and female at the same time, called *hermaphroditism.* The mangrove-dwelling killifish, *Rivulus marmoratus,* has a functional *ovotestis* that enables it to fertilize itself and produce offspring—natural cloning. In other hermaphroditic fish, cross-fertilization between different individuals occurs during spawning, but *Rivulus* always reproduces alone. It is the only known self-fertilizing hermaphroditic vertebrate.

SIX

Dangers and Defenses

Question 1: What diseases are common among fishes?

Answer: Fish, like humans, are subject to all sorts of diseases. But since there are thousands of different species of fishes, they have many more diseases, most of which we know nothing about since they haven't been studied. Fish disease is particularly important to aquarium hobbyists and to aquaculture operations, so diseases of those species have been the subject of considerable study.

Gas bubble disease (GBD) results from supersaturation of gases in the water. It can occur in aquarium or aquaculture fishes in water that was pumped under pressure, or in wild fishes living in water below dams, where entrained air is dissolved in water under pressure from the deep plunge. Gas bubbles form mostly in the blood, gills, eyes, and under the skin. Bubbles within capillaries restrict blood flow and form hemorrhages and clots. The effects of GBD, analogous to "the bends" in human divers, can range from mild to fatal depending on level of gases, fish species, life-cycle stage, depth, and such. Fish may have bubbles visible under their skin or show signs of swimming upside down or vertically.

Some diseases that have been investigated in wild fishes appear to be associated with water pollution. Tumors are found more frequently in highly polluted areas; for example, Dover sole, *Microstomus pacificus,* in urban estuaries have liver tumors. Tumor surveys can be useful in indicating the presence of

Common Infectious Diseases in Aquarium Fishes

"Ich," "Ick," or White Spot (*Ichthyophtirius*): Small white spots caused by a protozoan, mainly on the fins. When the protozoa first attach themselves to the fish, they are not visible to the naked eye, but they become visible after feeding on the body fluids of the fish for a few days.

Fin Rot: Rotting fins and loss of appetite due to bacteria that infects the fins, which may be susceptible due to fin nipping by other fish or to poor water quality.

Fish Fungus: White or gray cotton-like growths on the body. Fish seem to be more likely to have fungal rather than bacterial infections. Of the many different types of fungi that infect fish, *Saprolegnia*, the most common, is a filamentous fungus that appears as gray or white patches. It establishes in a specific area then spreads rapidly over the body and gills. Fungi often infect fish secondarily after some other infection, stress, or injury. Fungi also affect fish eggs, initially unfertile ones, but fertile eggs as well.

Pop-Eye: One or both eyes appear to be protruding abnormally due to a bacterial infection. The eye bulges out because of fluid buildup behind or in the eye itself. The bulging eye may have a thin layer of tissue around it that covers it, keeping it in the socket. The fish may also be less active and uninterested in food. Pop-eye can also be caused by gas bubble disease, resulting from excess levels of gas in the water.

Swim Bladder Disease: Fish have a difficult time staying upright and may hang in the water (see chapter 3, question 5: "How do fishes prevent themselves from sinking?").

Dropsy: Fluids build up and cause the fish to bloat and scales to protrude due to a bacterial kidney infection, causing fluid accumulation or kidney failure.

Velvet or Rust: A protozoan disease producing yellow to light brown "dust" on body, clamped fins, and respiratory distress. The fish may show signs of irritation, like rubbing against objects. The gills are usually the first part affected. It is highly contagious and can be fatal.

Common Infectious Diseases in Aquarium Fishes

Lymphocystis: A viral disease causing nodular white swellings (cauliflower) on fins or body. It can be infectious, but is usually not fatal.

Tumors: Tumors can be caused by a virus, environmental factors, or may be genetic. This is a common problem for professional breeders as the genetic tumors may result from too much hybridization. Certain hybrids of platyfish and swordtails invariably develop melanoma, a pigment cell tumor.

cancer-causing pollutants in the water or sediments. High frequencies of liver and skin cancers in brown bullheads, *Ameiurus (Ictalurus) nebulosus,* in the Black River, Ohio, and the Buffalo River, New York, were associated with high concentrations of polynuclear aromatic hydrocarbons (PAHs) and metals in the sediment. Similar pollutants and tumors are seen in English sole, *Parophrys vetulus,* in Puget Sound near Seattle, Washington.

Anglers may note some disease in a fish and wonder if it is safe to eat. The safest response is, "When in doubt—don't." However, very few fish diseases can be transferred to humans. Almost all fish are safe to eat when thoroughly cooked, smoked, or frozen. However, a tumor may be a response to pollutants, so the fish may have elevated levels of cancer-causing chemicals in its tissues.

Question 2: What kinds of parasites do fishes have?

Answer: A fish can be considered a swimming community of parasites. Parasites are typically divided into ectoparasites, which live on the outside (including gills, mouth, skin, and fins), and endoparasites, which live in the blood or organs (including intestines). Some parasites, including most ectoparasites, are transmitted directly from fish to fish, but they often involve a free-living phase in the water. This is called a direct life cycle.

Other parasites may have complex indirect life cycles involving two or more hosts.

Fishes may harbor thousands of internal parasites, including protists (single-celled ciliates, flagellates, microsporida), spiny-headed worms (acanthocephala), flukes (trematodes), tapeworms (cestodes), and roundworms (nematodes), as well as external parasites like isopods and copepods (crustaceans), and *glochidia,* the larvae of freshwater mussels. Microscopic glochidia have hooks to attach to the gills of a fish for a period of time before they detach, fall to the bottom, and mature into juvenile mussels. Since a fish is free-swimming, this helps distribute the mussels to areas they could not reach otherwise. While glochidia are called "parasitic," they do not appear to harm the fish.

Some internal parasites have complicated life cycles with different stages living in different hosts; the fish may be the final host for the adult (reproductive) stage or may be an intermediate host. Some fishes may have intermediate stages of parasites that can be transmitted to humans that eat them. A typical life cycle for trematodes is to have their first stage in a snail, second stage in a fish, and adult stage in a bird. Birds release the fluke's eggs in their droppings into the water, which are taken up by snails. The eggs hatch in the snail, and the next stage leaves and seeks their next host, a fish. Latching onto the fish's gills, they may remain there or move to other parts of the body and wait for the fish to be eaten by a shorebird. The common stage in fish is the *metacercaria,* which forms cysts in tissues and rarely causes major problems. When the fish is eaten, the parasites break out in the birds' stomach and move into the intestines, sharing its food and laying eggs that leave through the bird droppings— where the cycle begins again.

Kevin Lafferty and A. Morris of the U.S. Geological Survey found that parasites can actually alter a fish's behavior to make it more conspicuous to birds. California killifish, *Fundulus parvipinnis,* infected by the trematode *Euhaplorchis californiensis* may have thousands of metacercaria packed in their braincase. While fish normally evade birds by swimming to the bottom and remaining still, infected fish swim closer to the surface with shimmery, jerky movements that can attract bird predators. So this parasite changes the host's behavior in order to increase its chances of reaching the next host. While those parasites were in the fish's brain, where affecting behavior seems likely, Celine Santiago Bass at Rutgers University found similar behavioral changes in common killifish, *Fundulus heteroclitus*, infected with thousands of metacercaria in their gills.

Aquarium fishes frequently harbor parasites, as crowding in a tank can increase transmission of parasites—a major concern for hobbyists. There are chemical remedies available to kill the parasites without harming the hosts. Parasites are also a concern for aquaculture (where fish are also crowded), where chemical treatments such as antibiotics, copper, salt, and formalin are used to control parasites.

Crustaceans, particularly isopods and copepods, are common ectoparasites, sometimes called "fish lice." When numerous, they can debilitate the host, who might then seek out a "cleaner fish" if there are some in the neighborhood (see chapter 4, question 12: How do fish interact with other species of fishes?). Sea lice (copepods) from farmed salmon (see color plate G) can spread from fish farms to the young of nearby wild salmon populations (see chapter 9, question 12: Is aquaculture a possible substitute for commercial fishing?).

One ectoparasite that is particularly bizarre is the isopod *Cymothoa exigua*, which attaches itself to the base of a fish's tongue and extracts blood. As the parasite grows, less blood reaches the tongue, which eventually atrophies. The parasite then replaces the fish's tongue by attaching its own body to the tongue stub. The fish uses the parasite just like a normal tongue. This is a weird case of a parasite functionally replacing a host's organ.

Question 3: Can fishes themselves be parasites?

Answer: A few fishes are considered parasites. The sea lamprey is a jawless fish (see color plate A) with a permanently open, tooth-filled mouth at the front of its eel-like body. It fastens onto its prey and rasps out a hole with its rough tongue, leaving a rounded scar. An anticoagulant in the saliva keeps the wound open for hours, days, or weeks, until the lamprey is sated or the host dies. The site of attack, time of year, and size of lamprey and host all determine whether or not the host survives. When they first arrived in the Great Lakes, lampreys killed large numbers of sport fish—one lamprey may kill a dozen lake trout during its one- or two-year stay. Lampreys begin life as burrowing freshwater larvae that eat microorganisms for five to seven years. They transform into adults in a striking metamorphosis and become efficient swimming parasite-predators that typically migrate to the sea. When they are ready to reproduce, lampreys return to fresh water, build a nest, spawn, and die.

Pearlfish, in the genus *Carapus,* typically live in the gut of shallow-water sea cucumbers or the mantle cavity of clams. They enter the cucumber's anus tail first and may partly protrude from it. The fish generally leaves its host at night to feed on small fish and benthic (bottom-dwelling) invertebrates. Despite living inside a host, pearlfish do not appear to do harm. P. Glynn and colleagues from the University of Miami suggest that they are not really parasites, since they emerge at night to feed. This type of relationship is called *commensal*—it benefits the fish (by allowing them to avoid predators during the day) without harming the host.

The candirú, *Vandellia cirrhosa,* is a one- to two-inch-long eel-shaped catfish found in the Amazon and Orinoco rivers. It is almost completely transparent, making it nearly impossible to see. A fast, strong swimmer, it is smooth and slimy, with sharp teeth and backward-pointing spines on its gill covers. Among Amazonian natives, it is the most feared fish. To find a host, the candirú first tastes the water to locate a water stream coming from the gills of a fish. It follows the stream and inserts itself inside the gill flap. Spines around its head anchor it in place as it rasps the gills with its long teeth and slurps up blood. When finished, it unhooks its fins and sinks to the bottom of the river to digest its meal. The blood feeding has earned it the name of "vampire fish." However, the candirú is so feared because it is also attracted to humans when they are skinny-dipping or urinating in the water. It tastes the urine, follows it back to its source, and swims up the urethra and lodges itself in the urinary tract with its spines. The candirú gorges itself on the blood and body tissue, its body expanding from the blood, preventing the human host from urinating. Because of its spines, it is almost impossible to remove the fish except through surgery. The danger for the person is from infection and shock resulting from having the spiny fish lodged in sensitive tissues. It is difficult to imagine how even the most agile of fishes could squirm into someone's penis during a brief dip in the water, and women may be more vulnerable than men. No wonder native Amazonians would rather run into piranhas!

Question 4: What predators do fishes have?

Answer: Fishes are food for bigger fishes and many other animals. Just about any predator that is larger can eat fishes, so that for a fish larva the chance of survival to adulthood is minuscule. Fish larvae and small juveniles are even eaten by aquatic insects like dragonfly larvae and beetles and by spiders. Cnidarians—jellyfish, sea anemones, et cetera—sting and capture small fishes. Squids, prawns, and crabs eat fishes. Vertebrates that eat fishes include other fishes, frogs, salamanders, snakes, turtles, and, especially, birds and mammals. Fish-eating birds may wade along the shore (herons), attack from the sky (ospreys), or outswim them in the water (penguins). Piscivorous mammals include minks, otters, seals, sea lions, and both small and large whales. There is even a fish-catching bat. Human beings are perhaps the most efficient predator on fishes, driving some to the brink of extinction.

In addition to these typical larger predators, a very tiny "predator" has been found that can kill and partially consume large numbers of fishes. *Pfiesteria* is a microscopic dinoflagellate responsible for harmful algal blooms in the 1980s and 1990s along the coast of North Carolina and Maryland in which fishes, mostly menhaden, were killed with open bloody sores on their bodies. *Pfiesteria* caused massive fish kills, including one in the Neuse River in 1995 that killed 14 million fishes and closed 364,000 acres of shellfish beds. Discovered by JoAnn Burkholder at North Carolina State University, *Pfiesteria piscicida* is reported to have a highly complex life cycle with twenty-four forms. The name *piscicida* means "fish-killer." This single-celled organism produces a toxin that kills fishes. However, the related *Pfiesteria shumwayae* kills fishes primarily by feeding directly on their skin. Only where fishes were in direct physical contact with it did the fish die, and so this species could be considered a predator.

Question 5: How do fishes protect themselves against predators?

Answer: Defenses against predators can be morphological or behavioral. Morphological defenses can include structures like armor and spines as well as camouflage and color change (see color plates D and F). Bioluminescence by potential prey fishes may cause a "confusion effect" in a predator and reduce predation. Being toxic (like some pufferfish), having thick armor coating (like trunkfishes), or having sharp spines are also effective deterrents to predation. Fish can also modify their anatomy when predators are present; they can become taller or develop thicker spines than individuals in other environments. This is called *phenotypic plasticity* if the change happens in the same individual, or natural selection (evolution) if it happens over generations. Threespine sticklebacks, *Gasterosteus aculeatus,* (see fig. 17, in chapter 5), in areas with predators develop thick armor plates that are missing in fish without predators. The amount of armoring can change rapidly, as seen by Catherine Peichel of Stanford University. When Lake Washington in Seattle, Washington, was highly polluted in the 1960s, the water was murky and sticklebacks had little need for bony armor to protect them. After the lake was cleaned up, its clarity increased, and over the past forty years about half of the sticklebacks have evolved to become fully armored, with bony plates protecting their bodies from head to tail, with others partly plated.

Schooling is one of the behavioral defenses of fishes (see color plate E and chapter 4, question 5: Why do some fishes swim in schools?). Predators seem to eat less when confronted with a confusing school of prey rather than individuals. Fishes can warn one another about a predator using alarm substances, sound, or visual signals, then quickly leave the area. A common strategy is to hide or to remain motionless, so as not to attract the predator's attention; another is to keep one's distance. Differences exist between populations that have experienced a particular predator and those that have not; some differences are innate, but learning also plays a role. Hatchery-reared fish that are later stocked in natural waters have greater mortality after release than wild-born fish since they had no previous experience with predators. The ability to discriminate between more dangerous and less dangerous predators is also important.

Prey face a tradeoff between antipredator behavior (e.g., hiding) and the benefits of other activities, such as feeding and reproduction. When they are constantly on the lookout for predators, they will not be able to eat as much and may grow more slowly. These are trade-offs between safety and feeding; hungrier fish will take greater risks. Researchers have found differences between individuals in a population and between populations in boldness and willingness to approach a predator. In some cases, trade-offs aren't needed: groups of some prey fish (butterflyfish, damselfish) will mob the predator (eel, lizardfish) and drive it away.

Fish populations can evolve life history differences rapidly depending on whether or not predators are present. Guppies were introduced to environments with and without predators. At the site with predators, after eight years (under thirty generations) females produced more embryos at each reproductive cycle, according to Swanne Gordon of the University of California, Riverside, and colleagues. This makes sense biologically, because where predators are present, one might not get another chance to reproduce. In safer habitats, females produced fewer but larger embryos each time, thus expending fewer resources on reproduction and producing babies that were more competitive.

Question 6: If a fin is injured, does it grow back?

Answer: The phenomenon of regenerating lost tissues and organs from adult, differentiated cells is fascinating. While most vertebrates, including humans, do not have much ability to regenerate, fish do it well. Teleosts can regenerate a variety of body parts, including fins, the part most likely to be injured. After the cut end heals, the cells at the tips of the fin rays undergo *dedifferentiation* to become like embryonic stem cells and form a regeneration bud, or *blastema*. The cells divide and produce new tissue, the fin slowly grows, the cells re-differentiate, and the fin returns to its original condition within a few weeks. The actual time depends on the extent of the injury.

Question 7: What dangers do fishes face from people?

Answer: Probably the greatest danger fishes face from people is fishing, as discussed in chapters 8 and 9. There are few areas around the world without fishing. Other dangers arise from the release of foreign (non-indigenous) fishes, discussed later in this chapter. Yet other dangers arise from two additional kinds of human activities: habitat alteration or destruction and pollution.

Many human activities alter or destroy fish habitat in inland waters, estuaries, and the ocean. These include dams, which are a major barrier to migration and alter the river above it into a lake. Many fishes need flowing water, and the muddy bottom in the impounded water behind the dam will not meet their needs. Any human activities that alter the nature of the bottom (rocks, woody debris) can harm fishes. Other habitat changes occur with the building of sea walls or bulkheads that change the gradual slope of the beach into a vertical wall, removing shallow water habitat that is used by small fishes to avoid predation by larger ones. Other coastal alterations include filling in of tidal marshes or mangrove swamps to develop towns and cities. Many fish depend on tidal marshes for food and/or shelter, either for their whole lives or during juvenile stages. Channelizing a streambed (bank stabilization) involves straightening out the sides of a river, basically turning it into a pipe. This alters water flow and harms fishes that need the water that moves slowly. Logging forests near salmon-spawning streams causes silt to enter the water, settle on eggs, and impair development. Without trees, the water becomes warmer and less suitable for salmon. Coral reefs contain the most diverse fishes and are being destroyed by coral collecting, boat anchoring, harmful fishing techniques such as dynamite fishing, as well as by climate change, all of which result in fewer and less diverse fishes.

The Atlantic States Marine Fisheries Commission in 2009 released a report about the status of diadromous fishes (migrating between oceans and rivers) in which habitat alterations—dams, channelization, and land use—were considered major threats to herrings, shad, sturgeon, striped bass, and eels. Other major threats were toxic and thermal discharges, atmospheric deposition, low dissolved oxygen, and global warming, which are aspects of pollution. Power plants may draw in river water for cooling and then discharge warmer water back to the river. A plant can use billions of gallons a day and kill billions of fishes that are trapped on the intake screens (bigger fish) or drawn through the cooling system and cooked (smaller ones).

Pollution can have both subtle and not-so-subtle effects. Sewage and fertilizers entering the water from urban and agricultural regions act as nutrients that stimulate algal blooms (*eutrophication*) that can cause oxygen depletion, especially in the late summer, killing fish that are unable to move elsewhere or to breathe air. Some fishes can breathe at the surface, where the water holds more oxygen, but this makes them more visible to predators like birds. Because of the vast agricultural areas that drain into the Mississippi River, there is a "dead zone" that forms every summer in the Gulf of Mexico that is often the size of New Jersey. Dead zones are increasing and will be worsened by global warming, which will make more areas unfit for fishes. Excess nutrients also promote harmful algal blooms (HABs) that are toxic to marine life. Blooms of *Karenia brevis*, commonly referred to as "Florida red tides," occur regularly along Florida's west coast, causing massive fish kills. The term "red tide" is not accurate since there are many highly toxic cells that don't discolor the water, so "harmful algal bloom" is a more appropriate term. HABs of many species occur throughout the world and may play an important role in affecting fish communities. Larval fish are especially sensitive, as Jennifer Samson found, working in my lab at Rutgers.

Toxic chemicals (including metals, pesticides, detergents, and petroleum products) enter the water from mining, industry, oil spills, and agriculture. Some of the most harmful pesticides like DDT and industrial chemicals like PCBs (polychlorinated biphenyls) have been banned, but since they are very stable (persistent) and do not break down to innocuous chemicals for a very long time, they are still cycling through aquatic sediments and food chains. They *bioaccumulate,* or concentrate in animals as well as sediments. Some pollutants, like DDT-related pesticides, PCBs, and mercury, *biomagnify* through food chains—at each step in a food chain, the predator concentrates the chemicals from its prey—so that the largest predatory fishes acquire the highest concentrations and are most likely to be affected. They are also able to pass this on to their predators (people), and the public is advised to restrict consumption of certain fishes because of contaminants (see chapter 10: "Fish and Human Health"). Direct release of many chemicals has been restricted, their levels have been reduced, and direct kills of fishes are now relatively rare, although many streams near mines remain devoid of fishes. However, lower concentrations of toxic chemicals have sublethal effects, impairing reproduction, behavior (e.g., activity, schooling, prey capture, predator avoidance), embryonic development, growth, and disease resistance. Some chemicals alter the endocrine system at very low concentrations, affecting hormones and, therefore, many aspects of a fish's life. John Sumpter of Brunel University in England has found that pharmaceuticals from human urine pass through sewage treatment plants unchanged and affect fish. Many fishes near sewage outfalls are intersex or otherwise reproductively abnormal due to estrogens from birth control pills.

Climate change can have enormous effects on fishes; it can destroy corals, which are important habitat. Excess carbon dioxide dissolved in the water increases acidity and reduces shell deposition for corals and other shell-forming animals. Warmer temperatures cause corals to lose the symbiotic algae (bleaching) that are responsible for photosynthesis that powers the reef. This can result in death of corals. Warming temperatures also affect fishes directly and reduce habitat for species dependent on cold water. In some species the temperature during embryonic development determines the sex so that warmer temperature will affect the sex ratio.

Question 8: What dangers do people face from fishes?

Answer: When the topic is dangerous fishes, most people think about sharks. They certainly have had a lot of bad press, thanks in part to *Jaws* by Peter Benchley, movie by Steven Spielberg. Some sharks are large and seek large prey, occasionally humans, but the hysteria that arises in the public is unwarranted. The species most likely to attack humans are the great white shark, *Carcharodon carcharias;* the tiger shark, *Galeocerdo cuvieri;* mako, *Isurus;* bull shark, *Carcharhinus leucas;* and hammerhead, family Sphyrnidae. Weighing over two tons and as long as a bus, the great white is the most famous marine predator. There are approximately one hundred reported cases of shark attacks and about five fatalities each year worldwide, though the true figure may be higher since records in some countries may be unreliable. However, this figure is still far less than deaths due to other dangerous animals such as bees (by stings), scorpions, or crocodiles. Spear fishermen with captured fishes and surfers lying on their boards appear to be likely targets. It is possible that, to a shark, the surfboard with arms and legs sticking out looks like a seal. Surfers splash a lot too, making it easier for sharks to mistake their identity. Since 1980 over three hundred surfers worldwide have been mauled by sharks. Shark attacks appear to be increasing near places with shark-feeding tours or resorts that feed sharks to keep them in the vicinity. If the sharks learn

Figure 21.   Great barracuda, *Sphyraena barracuda,* dining. *(Courtesy of NOAA.)*

that a free meal is available in a certain area, they are likely to associate that location with food and may mistake a human for food.

Other fish that will attack humans are the barracuda, *Sphyraena barracuda* (fig. 21), with their knife-like teeth. They appear to be attracted to shiny objects and bright colors. However, attacks are quite rare, and these fish are generally rather placid unless provoked. The highest incidence has been on swimmers splashing at the surface or swimming in turbid water. Because barracudas are visual predators, mistaken identity is probably involved. Moray eels (see color plate H) are secretive and live in caves and crevasses. If people stick their hands in or harass them they are likely to respond in an unfriendly manner, but unprovoked attacks are almost unheard of. One of the most feared fish are the small South American piranhas (see fig. 13, in chapter 4), with razor-sharp teeth, that can attack and kill animals much larger than themselves (including people) by working in groups.

A

Sea lamprey, *Petromyzon marinus*, mouth (oral disc) view. *(Courtesy of U.S. Environmental Protection Agency.)*

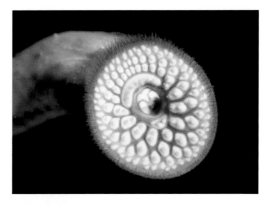

Whale shark, *Rhincodon typus*. *(Photo by Zac Wolf, Wikimedia.)*

Mudskipper, *Periophthalmodon septemradiatus*. *(Photo by Gianluca Polgar.)*

Hawaiian triggerfish, *Rhinecanthus rectangulus*. *(Photo by Larry Winnik.)*

Rockfish, *Sebastes* sp. *(Courtesy of NOAA.)*

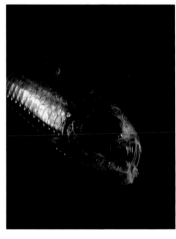

Luminescent deep-sea viper-fish, *Chauliodus sloani. (Photo by David Shale.)*

Coral reef scene with a school of longfin bannerfish, *Heniochus acuminatus,* with long dorsal fins, that may be confused with Moorish idols. *(Courtesy of NOAA.)*

Moorish idol, *Zanclus cornutus. (Photo by Mila Zinkkova, Wikimedia.)*

C

Lionfish, *Pterois volitans.* (Photo by Tobias Biehl, Wikimedia.)

Blind cave fish, *Astyanax mexicanus,* with sighted offspring. Note absence of pigmentation as well as eyes. (Photo by Richard Borowsky.)

Sockeye salmon, *Oncorhynchus nerka,* arrive at spawning grounds. (Courtesy of NOAA.)

Rainbow trout, *Oncorhynchus mykiss.* (Courtesy of U.S. Bureau of Reclamation.)

D

Peacock flounder, *Bothus lunatus*, master of disguise. *(Photo by Peddrick Weis.)*

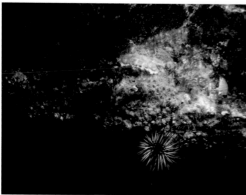

Toadfish, *Opsanus beta*, is also good at camouflage. *(Photo by Gregg Carter.)*

Seahorse, *Hippocampus* sp. *(Courtesy of NOAA.)*

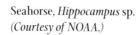

Longear sunfish, *Lepomis megalotis. (Courtesy of New York State Department of Environmental Conservation.)*

E

School of blue-striped grunts, *Haemulon sciurus*. *(Photo by Jeroen Valkier.)*

Two cleaner wrasses, *Labroides phthiorphagus*, work on a needlefish, while two convict tangs, *Acanthurus triostegus*, await their turn. *(Photo by Mila Zinkova, Wikimedia.)*

Clownfish, *Amphiprion* sp., at home in anemone. *(Photo by Jeroen Valkier.)*

Redband parrotfish, *Sparisoma aurofrenatum*. *(Photo by Deron Burkepile.)*

F

Leafy sea dragon, *Phycodurus eques.* (*Photo by Dro!d, Wikimedia.*)

Swordtail, *Xiphophorus helleri.* (*Photo by rion819, iStockphoto.*)

Siamese fighting fish, *Betta splendens,* with bubble nest. (*Photo by ergoSum88, Wikimedia.*)

Discus, *Symphysodon* sp. *(Photo by Anka Zolnierzak, Wikimedia.)*

Bluehead wrasse, *Thalassoma bifasciatum.* *(Photo by Peddrick Weis.)*

Pink salmon fry, *Oncorhynchus gorbuscha*, with sea lice parasites. *(Photo by Alexandra Morton.)*

H

Moray eel, *Gymnothorax javanicus*, with hump-back cleaner shrimp (*Lymata*) dental hygienist. *(Photo by Michael Schmale.)*

Scorpionfish, family Scorpaenidae. *(Photo by Craig Kasper.)*

Spiny pufferfish, juvenile. *(Courtesy of NOAA.)*

Mosaic of loaves and fishes in the Church of the Multiplication, in Tabgha, Israel. *(Photo by Peddrick Weis.)*

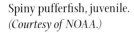

People can also be harmed by shocks from electric eels, *Electrophorus;* rays, *Torpedo;* and catfishes, *Malapterus*. The voltage can cause pain and temporary numbness (see chapter 3, question 16: How and why do some fishes produce electricity?). Billfishes and sawfishes have a long, pointed extension in the front that can penetrate wooden boats and injure those in the boat or water. Needlefishes, Belonidae, with sharp, pointed beaks may strike swimmers or surfers and cause serious damage.

A number of fishes have venomous spines that can inflict painful stings that combine physical injury with release of venom. The danger from stingrays is well known (especially after the freak accident that took the life of "crocodile hunter" Steve Irwin); however, they seldom injure people that seriously. The stinging spines, located at the base of the tail, are about 6 inches long, with a groove. Venom is produced in a gland in the groove, and when the spine is embedded, it tears off and remains in the wound. Attacks usually result from someone stepping on the bottom-dwelling ray, which whips its tail around, sticking spines into the person, who would be wounded in the leg and have a serious and painful wound, but not die, as Irwin did because the spine penetrated his heart. Symptoms include a drop in blood pressure, nausea, faintness, dizziness, and shock. Some other cartilaginous fishes also have venomous spines, as do some teleosts, including surgeonfishes, scorpionfishes (see color plate H), catfishes, and several other groups. Some catfishes (including some in home aquaria) have venom glands in the skin near the dorsal and pectoral spines; their stings are painful but not dangerous. They should be handled with care and grasped behind the pectoral fins to reduce the danger of being stung. Scorpionfishes, most common in tropical shores and coral reefs in the Indo-Pacific, tend to have venom glands in grooves along the dorsal, anal, and pelvic spines. The family is large, with hundreds of members. In this family is the colorful lionfish, *Pterois* (see cover and color plate C) which has very potent venom and has killed humans. This Indo-Pacific native is sometimes kept in home aquaria.

There is also a possible risk to people of catching a *Mycobacterium marinum* infection from handling fishes. "Fish TB" produces skin infections that heal slowly; most go away on their own after a few months. One cannot get "lung TB," *M. tuberculosis,* from a fish. Since Mycobacteria are found in aquarium fish and there is a low incidence of cases in humans, the risk of acquiring this infection is low. It is more common among fish market employees, whose hands are frequently nicked during scaling and gutting fish, allowing entry of the bacterium. This has become less common where regulations require the wearing of disposable gloves.

Question 9: What problems arise when non-native fish are released in local waters?

Answer: Fish may be deliberately relocated for aquaculture and for recreational fishing or relocated accidentally—perhaps discarded after being caught or purchased elsewhere or as stowaways in ships' ballast water. When fish arrive in a new location, they are usually harmless or do not survive. But sometimes they do not integrate well and cause costly ecological and economic problems. Those that thrive may become invasive by outcompeting, eating, or parasitizing native species. The introduction of the predatory Nile perch into Lake Victoria, East Africa, led to the extinction of over two hundred species of native cichlids as well as other native fishes, according to John Balirwa from the National Fisheries Research Institute of Uganda.

One of the first aquatic invasive species (not a fish) to receive a lot of attention in recent years was the zebra mussel, *Dreissena polymorpha,* that arrived in the Great Lakes in the late 1980s, probably as a result of ballast water discharge from ships from eastern Europe. In its new location, this mollusk had no natural predators and grew very densely, completely covering hard surfaces. They clogged up intake pipes for water systems, causing great economic and ecological harm. Several years later, a fish called the round goby, *Neogobius melanostomus,* arrived, probably also in ballast water. The good news was that these fish, unlike any native to the Great Lakes, eat zebra mussels, almost eighty a day. The bad news is that gobies can displace native bottom-dwelling fish. Major reductions in populations of sculpins have occurred in areas where gobies have become established because gobies compete with sculpins for food or drive them from their habitat and spawning area. Gobies also eat darters and other small fish and feed on eggs and young of lake trout, which already have reduced populations in the Great Lakes. By changing food webs, gobies have increased the potential of contaminants to reach humans. Zebra mussels accumulate pollutants, which can be transferred to the gobies that eat them. This is of concern because gobies are eaten by sport fish (smallmouth and rock bass, walleyes, yellow perch, and brown trout), which accumulate contaminants and transfer them to human consumers.

Asian carps, ravenous and rapidly breeding, have been multiplying in the Illinois River and threaten the well-being of native fish and commercial fishing. In the early 1970s, bighead carp, *Hypophthalmichthys nobilis,* and silver carp, *Hypophthalmichthys molitrix,* were imported from the Far East to eat algae in fish farms in the South. But flooding helped them to escape. By 2003 they had gotten out of control; that river may have the largest population of Asian carp in the world, according to Greg Sass of the Illinois Natural History Survey. Carp frequently jump out of the water into boats, causing injury to passengers, according to the *New York Times.* An electric fence is in the Illinois River to stop them from getting into the Great Lakes. In addition, agencies poisoned a stretch of the Chicago Sanitary Ship Canal to kill off the fish while maintenance was performed on the electrical barrier, and also considered temporarily closing the vital shipping corridor to stop the spread of the fish into Lake Michigan. DNA evidence suggests that they had already entered Lake Michigan by late 2009.

Invasive species in the Pacific Northwest pose a threat to native salmon by predation. Beth Sanderson of the Northwest Fisheries Science Center and colleagues identified six nonindigenous fishes—catfish, black and white crappie, largemouth bass, small-

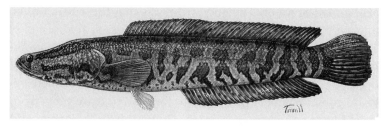

Figure 22. Northern snakehead, *Channa argus. (Courtesy of Michigan Department of Environmental Quality.)*

mouth bass, walleye, and yellow perch—that were consuming hundreds of thousands to millions of endangered juvenile salmon at just a few sites. Salmon constituted a large fraction of the diet of some of the invasive species. This appears to be a major source of mortality, comparable to better-known threats such as fishing, hatcheries, dams, and habitat alteration.

During the summer of 2002, exotic fish called the Northern snakehead (fig. 22) were found in a pond in Maryland, near Washington, D.C. A man had ordered them from New York's Chinatown to prepare a traditional soup remedy for his ill sister. However, by the time the snakeheads arrived she had recovered, so he released them into a local pond. The potential impact was considered so damaging that the event made national headlines. By the time they eradicated them from the pond, the fish showed up in the Potomac River and established a population. The Northern snakehead, *Channa argus,* is a voracious predator with no natural enemies. Most of its diet is other fish, though it also eats crustaceans, insects, and plants. It can live in water from 32 to 87 degrees Fahrenheit and is found in muddy or vegetated ponds, swamps, and slow-moving streams. They can breathe air and survive up to four days out of water, which enables them to travel over land to new bodies of water by wriggling over the ground. These adaptations to the seasonal drying of shallow ponds in their native habitat in China allow them to disperse widely. They are sold in the United States both as food in Asian markets and as pets, prized for hardiness and aggressive habits. Many snakeheads that appear in the United

States are thought to be former pets that were released when their owners no longer wanted them. Reproducing populations have now been discovered in Maryland, California, and Florida, and individuals have been caught in Maine, Massachusetts, Rhode Island, New York, and Hawaii. Scientists are concerned that established populations will have disastrous consequences for their new habitat.

Another way species get to new locations is through ship canals and locks that connect the Atlantic Ocean with the Great Lakes, bypassing obstacles like Niagara Falls. An unintended consequence of these canals has been the introduction of the sea lamprey *Petromyzon marinus* (see color plate A) to the Great Lakes. It has been very damaging, feeding parasitically on host fish like lake trout. But there is good news. It is one of the few aquatic invasive species that is being successfully controlled. A vulnerable point in their life cycle is the larval stage, when they spend three years burrowed in stream sediments. A chemical, TFM (3-trifluoromethyl-4-nitrophenol), is selectively toxic to lampreys and has been carefully applied to infested streams, decreasing lamprey numbers by 90 percent and allowing native lake trout to survive and rebound. A new approach to lamprey control is being developed by Nicholas Johnson and Weiming Li of Michigan State University, utilizing a lamprey-specific sex attractant (pheromone) rather than a synthetic poison like TFM. They released the synthetic version of a male pheromone from traps placed in a stream. Ovulating females smelling it swam vigorously upstream until they found the source, some becoming trapped in the process.

Red lionfish, *Pterois volitans* (see cover and color plate C), a venomous species from the western Pacific Ocean, have been found in the past decade along the Atlantic Coast from the Carolinas to the Caribbean; small numbers have been found as far north as Long Island. With maroon and white stripes and a plume of feathery spines, they are popular aquarium fish. Without natural predators, they compete with local fish for space and food (including smaller fish that they eat). They were probably introduced accidentally in the early 1990s, perhaps escaping from

Florida tanks during flooding from Hurricane Andrew in 1992. With no natural predators, they are thriving and putting native species at risk; they also are a menace to divers and anglers who may come in contact and be stung. In the Bahamas they are voraciously eating most of the reef fishes, including commercially important species. Efforts are underway to educate anglers about how to catch them, clean them, and eat them safely.

Invasion of aquatic ecosystems does not only bring alien fishes to the United States. Some of our natives have invaded other parts of the world. The rainbow trout, *Oncorhynchus mykiss*, beloved of fly fishers (see color plate C), has been introduced for food or sport in over forty countries. In southern Europe, Australia, and South America, they have had serious negative impacts on native freshwater fish by eating them, outcompeting, transmitting diseases, or hybridizing with closely related species. The mummichog, *Fundulus heteroclitus*, a resident of our salt marshes (see fig.16, in chapter 4), arrived in Spain in the 1970s and became established in the 1980s. It is not known how they got there, but ballast water is a likely route. In the Ebro River delta, they are now the most abundant fish in the marshes. Their small size, short life span, early maturity, and high tolerance for stressful conditions like low oxygen and high levels of pollutants make them very successful. Sharing the marsh habitats with the invaders is a native species in the same family, *Aphanius baeticus*, which is endangered. This species is likely to have an even tougher time surviving with mummichogs taking over their habitat.

SEVEN

Watching Fishes

Question 1: How popular is fish keeping as a hobby?

Answer: People like to watch fish. The popularity of fish screen-savers attests to that. Fish keeping has a long history, going back over two thousand years. Depictions of fishes kept in temple pools have been found in ancient Egyptian art. Many other cultures have a history of keeping fishes for decorative purposes. Fish keeping is now a multimillion-dollar industry, with the United States being the largest market in the world. While more than half of the households in the United States have pets, over 10 percent own ornamental fishes. The vast majority of these fishes are in indoor aquaria, while a small number, mostly decorative goldfish or koi, live in outdoor ponds which are much larger and exposed to the elements. While there is some disagreement over whether fishes are pets or a hobby, there are hundreds of clubs devoted to tropical fishes in general or to a particular species and dozens of newsletters and magazines devoted to fishes—how to keep them, how to breed them, how to set up a tank and maintain good water quality, and so forth. Fish fans and fanatics have numerous blogs and Web sites so that anyone can ask questions and learn from experts.

Watching Fish in Art

Fishes have been represented in art from ancient times to the present. From ancient mosaics in Israel (see color plate H) to ancient Mexican and Northwestern Indian tribal drawings, to ancient Roman and Medieval images, fishes have been a constant design in Western art. In Japan and China, fishes have been an important theme in art for centuries. With the development of printing, the non-religious depiction of fishes became more widespread, and realistic paintings of fishes appeared in still-life paintings during the Renaissance. In the twentieth century, fishes were painted by many modern artists, including Henri Matisse, Pablo Picasso, and Paul Klee. Fishes are not only the subject of art, but a medium as well, where the fish is used as a printing block. Fish prints (gyotaku, a traditional Japanese art) are made by applying water-based inks directly onto the side of a real fish and pressing rice paper on top. When lifted, a detailed mirror-image fish print is seen. Even though more than one print can be made from the same fish, they will not be identical.

Question 2: What are the pros and cons of freshwater and saltwater community tanks?

Answer: Home aquaria are popular ways to become familiar with fishes and their behaviors. Freshwater tanks are the most popular—they are cheaper and easier to maintain. Marine aquaria generally require more complex equipment and are more expensive to set up and more difficult to maintain. The fishes are also more expensive. Marine aquaria frequently feature a diverse range of invertebrates such as shrimps and anemones. Certain subtypes of aquaria also exist, such as the reef aquarium with corals and reef fishes. These can be exceedingly beautiful due to the attractive colors and shapes of the corals and reef fishes. Green aquatic plants are often used in both marine and freshwater aquaria since, through photosynthesis, they utilize waste carbon dioxide from the animals' respiration and in turn provide oxygen. Brackish water aquaria combine elements of both marine and fresh water. Fish kept in brackish water aquaria generally come from habitats with varying salinity, such as mangroves and estuaries. It is possible to collect water and fishes such as killifish from local estuaries and set up a brackish water aquarium, rather than buying the fishes. When you consider setting up a tropical aquarium, get a good reference book. You should learn about the species you want before you purchase it and consider how large it will get. For any aquarium, make sure that food for the species you have is consistently available. To keep your tank healthy, don't overfeed or overcrowd the fishes. You should test the pH, ammonia, and nitrates regularly, change 25 to 50 percent of the water at least once a month, and have a quarantine aquarium for new or sick animals. When the water evaporates, refill with fresh, unchlorinated water. Any change in the salinity or temperature should be made slowly.

Question 3: Which species adapt well and live peacefully in a home fish tank?

Answer: The community tank is the most common type of aquarium, where several non-aggressive species are housed peacefully together. The fishes, invertebrates (when present), and plants in the tank may not originate from the same geographic region, but tolerate similar conditions. Some of the most popular freshwater species that adapt and live peacefully in a community tank include the following:

Live-bearers: guppy, platy, sailfin molly, swordtail (see color plate F), and black molly

Catfish: bristlenose catfish, banjo catfish, bronze cory, Sterbai's cory, glass catfish, and sailfin plecostomus

Cichlids: angelfish, convict cichlid, discus (see color plate G), and kribensis

Tetras: neon tetra, black widow tetra, cardinal tetra, and
Congo tetra
Cyprinidae: White Cloud Mountain minnow, zebra danio,
rosy barb, and tinfoil barb
Cobitatae: clown loach and Kuhli loach

Of this mix, some will prefer to swim near the surface, some
will stay on the bottom and eat food that has fallen there, and
the rest will be somewhere in the middle, filling the gaps and
providing a balanced community.

Question 4: What are the sources of fishes for home aquaria?

Answer: Fishes can either be caught from the wild or be bred
in captivity. While over 90 percent of the freshwater aquarium
fishes are captive bred, virtually all marine aquarium fishes and
invertebrates are caught from the wild. In many developing
countries in South America and Asia, local villagers collect specimens for the aquarium trade as their prime means of income.
Over 200 million fishes comprising about 730 freshwater and
over 800 saltwater species worth over $44 million are imported
into the United States annually. Freshwater fishes account for
about 95 percent of the total fishes and about 80 percent of the
total value. The few marine species bred in captivity supplement
but rarely replace the trade in wild-caught specimens. Fishes
can be collected by net, trap, or cyanide. The most damaging
technique is cyanide, a poison used to stun fish to make them
easy to collect. However, it not only can damage and kill fishes,
but also kills corals. Once corals are dead, they are no longer
a good home for fish, and the reef degrades. Catching fish in
the wild can reduce population sizes, placing them in danger of
extinction in the collecting areas. Many fishes die during shipping; others are weakened by stress and become diseased upon
arrival. Dealers and hobbyists should avoid buying fish caught
by this method.

The negative impacts of fish collecting, including dynamiting or poisoning of coral reefs and non-target species, reduction of rare species from their natural habitat, and degradation of ecosystems from large-scale removal of key species, have come to the attention of aquarists worldwide. Many concerned people are trying to reduce the trade's dependence on wild-caught specimens through captive-breeding programs and certification programs for wild-caught fish. Among American aquarists surveyed, most prefer to buy farm-raised coral instead of wild-collected coral, and most said that only sustainably caught or captive-bred fish should be allowed for trade. Captive breeding for the aquarium trade has been growing rapidly since the mid-1990s, but only a few marine species are currently bred, including clownfish, some damselfish, and dwarf angelfish, all of which live well together provided there is enough space and niches for individual territories.

Question 5: What are the benefits of a public aquarium?

Answer: A public aquarium is like a zoo or museum and is a wonderful learning experience. Many facilities offer educational materials and programs, special rates for school groups, and customized tours. A good aquarium will have special exhibits in addition to its permanent collection. Some have their own version of a "petting zoo," a shallow tank with fishes and invertebrates that one can reach in to touch as they pass by.

The first public aquarium, called the Fish House, opened in the London Zoo in 1853. P. T. Barnum followed in 1856 with the first American aquarium as part of his Barnum's American Museum in New York. The oldest existing American aquarium is the National Aquarium in Washington, D.C., founded in 1873. This was followed by the opening of public aquaria in San Francisco, Woods Hole, New York, La Jolla, Detroit, Philadelphia, San Francisco, and Chicago. For years, the Shedd Aquarium in Chicago was the largest aquarium in the world, until the Georgia Aquarium opened in Atlanta in 2005. Modern aquarium tanks can hold millions of gallons of water and can house large species like dolphins, sharks, or beluga whales. This is accomplished though thick, clear acrylic windows. Aquatic and semi-aquatic mammals, including otters and seals, may also be on exhibit. Public aquaria may be affiliated with research institutions or conduct their own research programs, sometimes specializing in species and ecosystems from local waters. For example, the Vancouver Aquarium is a major center for marine research, conservation, and animal rehabilitation for the Pacific Northwest and MBARI (Monterey Bay Aquarium Research Institute) is a leading research institution.

Aquaria, like zoos and museums, have a unique potential to communicate conservation issues to the public. They have built-in audiences of millions of visitors, many scientific and other resources, and they are trusted sources of information. They may be the most important source of information to the ocean-conscious public, aquarium hobbyists, and industry operators (e.g., retailers) who visit them. Kids and adults can learn about many aspects of the marine environment, from species identification and habitat to conservation and preservation techniques. Aquaria can play an important role in raising awareness about conservation issues in the marine ornamentals trade and the need to buy organisms that have been raised or collected sustainably. By focusing on proper raising, collecting, shipping, and handling techniques, aquaria can enlighten the public about the oceans and their inhabitants.

Appendix 1 is a list, alphabetically organized by state, of many public aquaria in the United States. They all have Web sites with additional information.

There are public aquaria in other countries. Among the most impressive aquaria that I have visited is the one in Sydney, Australia. It contains a large variety of Australian aquatic life, over 650 species. Visitors walk through its see-through tunnels and watch sharks and other fishes gliding silently above. The diverse colorful fish and corals of the Great Barrier Reef are featured. Almost every Australian sea and major river is represented from

Watching Fish in Children's Literature

Fishes are common images in children's literature. A classic story, *The Rainbow Fish* by Marcus Pfister, is a lesson about the benefits of sharing. Dr. Seuss has *One Fish, Two Fish, Red Fish, Blue Fish*. Many other picture books for very young children, such as *How Many Fish?* by Caron Lee Cohen and S. D. Schindler, feature fishes for learning numbers and counting. Older children can read books such as *The Little Fish That Got Away* by Bernadine Cook and Crockett Johnson. Cartoons such as *Finding Nemo* and *Shark Tales* feature various fishes, and *Finding Nemo* is pretty accurate biologically (if you can accept talking fish).

the open sea to the Great Barrier Reef, to the Australian Bight and Sydney Harbor, to the Murray-Darling river system, rivers of the Far North, mangroves, and rocky shores. Opened in 1988, it is one of the largest aquaria in the world. Another impressive aquarium is in Eilat, Israel. The special feature here is the underwater observatory, which is below the surface of the Gulf of Aqaba of the Red Sea, so that the visitor is "inside" and the fishes are free, living in the surrounding coral reef. The observatory is twenty feet below sea level and is accessed via a pier from land and a staircase. The constructed reef provides a natural habitat that attracts many different kinds of fishes and invertebrates. Since the water in the gulf is generally crystal clear, observation of the diverse and beautiful fishes and corals is spectacular. The visitor can stay for hours watching the scene, which changes from window to window, and hour to hour. It is very unfortunate that in the construction of this unique observatory much of the local coral reef was destroyed. A similar undersea observatory is the Coral World in St. Thomas, U.S. Virgin Islands.

Question 6: What are the advantages and disadvantages of snorkeling versus scuba diving?

Answer: Snorkeling can be mastered quickly by anyone comfortable in the water and who can learn to breathe by mouth. You need to have a properly fitted face mask with a snorkel; fins are useful but not essential. If you have poor vision you can get a prescription face mask. You can lie quietly on the top of the water and watch the activity below. You can, if you wish, dive down to be closer to the animals while you hold your breath. Scuba (*self-contained underwater breathing apparatus*) requires lessons and certification. Trained divers can stay under the water for an hour or more, depending on depth, since air is used more quickly at greater depths. There are now rebreather systems that recirculate the inert gases and replace only the oxygen that is actually used. These allow divers to remain under water for many hours. Scuba requires wearing a heavy tank with compressed air on your back, a regulator, and a hose that connects to your mouth for breathing. You also have weights around your waist to help you descend and counteract the buoyancy of your body. The tank is connected to an inflatable vest for regulating your buoyancy—you can add air when you wish to be higher in the water and release air when you want to go deeper (just like a fish's swim bladder). Fins are necessary for scuba, and, of course, the face mask. Divers generally wear wet suits because the deeper water is colder, even in the tropics. Scuba burdens you with a lot of cumbersome gear (50 to 60 pounds worth), but once you are in the water and weightless you can see close-up what is going on, and you can remain still enough that the fishes are not disturbed.

Snuba is a trade name for a hybrid underwater-breathing system. The swimmer uses fins, a mask, weights, and breathing apparatus as in scuba. The air is not from heavy tanks on the back, but comes through a long hose from tanks on a boat at the surface. Snuba, also called a "hookah," often serves as an introduction to diving, in the presence of a professional. Since

Fish as Religious Symbols

Fish symbols are found in many of the world's religions. The Christian fish symbol is usually just two simple curved lines. Often seen on car bumpers, it goes back further than the cross as a Christian symbol. Christians began using the Greek word for "fish." *Ichthys* consists of five letters from the Greek alphabet: I-ch-th-y-s. When these five letters are used as initials for five words, they can spell out *Iesous Christos Theou Yios Soter,* an acrostic for "Jesus Christ, God's Son, Savior."

Fishes are mentioned and given symbolic meaning several times in the Gospels. Several of Jesus' twelve disciples were fishermen. He commissions them with the words "I will make you fishers of men." At the feeding of the multitudes, according to the Gospel of Mark, Jesus and his disciples had gone out in a boat in the Sea of Galilee, and crowds met them when they landed. It was dinnertime and they were not near anywhere to buy food. A boy brought Jesus five small loaves and two fishes, and Jesus miraculously multiplied them to feed the thousands of people. The fish was probably the Galilee tilapia, *Sarotherodon galilaeus,* or the blue tilapia, *Oreochromis aureus,* which are the most important food fish in that part of the world. This miracle is celebrated in the Church of the Multiplication in Tabgha, on the Sea of Galilee, in Israel, where there is a famous mosaic (see color plate H).

Fishes are also prevalent in Jewish tradition as food on the holiday table, especially on the New Year and Shavuot. They are a symbol of fertility because Jacob gave his children a blessing that they should multiply like fishes in the sea. Fishes are also associated with the coming of the Messiah; according to a legend, the Messiah will come in the form of a great fish from the sea. Gefilte fish, finely chopped fish, usually whitefish, pike, or carp, mixed with crumbs, eggs, and seasonings, cooked in a broth, and formed into oval-shaped cakes, is usually served chilled and is popular at Passover.

(continued)

Fish as Religious Symbols, *continued*

One of the eight auspicious symbols of Buddhism is a symbol consisting of two fishes, which appear standing vertically with heads toward each other. The pair of fishes originated as an ancient pre-Buddhist symbol of the two sacred rivers of India, Ganga and Yamuna. The golden fishes symbolize happiness, as they have freedom in water. They represent fertility and abundance as they multiply rapidly.

The Golden Carp is considered an auspicious symbol in ancient Eastern cultures. One legend tells that if the Golden Carp swims against the currents and leaps over the waterfall it will turn into a celestial dragon, one of the most honored and respected dragons in Eastern culture. It is also a symbol of wealth, prosperity, and success, and its likeness can be found throughout the East on textiles, plates, and murals. In Hinduism, the fish also appears in a sacred iconograph, the Avatars of Vishnu, where the deity is shown emerging from the mouth of a fish.

Last but not least, Pisces is one of the zodiac symbols, the twelfth astrological sign. Those born under this sign (February 20–March 20) are supposed to be full of feeling and empathy, emotions that are characterized by water. Pisces people are considered compassionate and loving but are not very rational and tend to avoid making hard decisions.

you are connected by the air hose to the boat, there is no concern about going too deep and you can choose the depth where you are most comfortable, down to about twenty feet. No matter whether you are watching from the shore, snorkeling, or diving, if the water is turbid or cloudy, go for a swim or do something else that day.

In all of the above methods of getting into the water, an underwater camera provides happy memories. Most of the photographs in this book were taken this way, some while snorkeling, others while scuba diving. The camera need not be expensive, and there are underwater housings available for many "point-and-shoot" digital cameras.

Question 7: How can I watch fishes underwater in their natural habitat if I can't swim?

Answer: Most people never imagine that they can look at fishes except in an aquarium or by snorkeling. But fish watching is possible. Go to a pond on a day when the water is calm, walk slowly along the banks and look down where the water is clear over light patches of sand or pale rocks. Look for shadows and bear in mind that the fishes may be quite small. Rings on the surface may be an indication of fish feeding. A bridge over a stream is a place where you can look right down into the water, and binoculars can be useful. For some observations you might want to wade into the water, but go slowly and carefully. It is difficult to see details and probably impossible to identify fishes if all you can see is the top, but if you wait, the fishes may roll on their sides, permitting you to see patterns. A difficulty in seeing into the water is surface glare, so you need to find a direction where that does not interfere with your viewing; polarized sunglasses "see through" glare. A glass-bottomed view box or "water telescope" will greatly improve what you can see. If you want to wade in the water, wear old sneakers or boots to protect your feet. Stay away from deep water—you won't see anything anyway. In colder weather you might want chest-high waders, but in deeper water losing your balance can be more than uncomfortable. At night many fishes move into shallow water and you can see them with a flashlight.

In many tropical areas tour operators run glass-bottom boat trips that go over a reef and permit observers to watch the colorful coral reef fishes while staying dry. For scientific expeditions, and to some degree for (rich) tourists, there are submersible vehicles that can stay under water many hours and can visit previously inaccessible deeper parts of the ocean.

EIGHT

Recreational Fishing

Question 1: What is recreational fishing?

Answer: Recreational fishing, or sport fishing, is fishing for pleasure or competition, as contrasted with commercial fishing (fishing for profit), or subsistence fishing (fishing for survival). The most common form of recreational fishing is with a rod and reel, but fishes can also be caught with minnow traps and nets. Spear fishing may be allowed for certain species in certain locations. The practice of catching fish with a hook and line, or angling, is easily done from a dock or a boat. Recreational fishing became popular with the publication of Izaak Walton's *The Compleat Angler* in 1653, a classic book for the angler who loves fishing for its own sake. Big-game fishing is fishing from boats to catch large open-water species such as tuna, sharks, and marlin. It became popular only after motorboats became readily available in the twentieth century.

Question 2: What is fly-fishing?

Answer: In fly-fishing, fishes are caught with artificial flies that are cast with a fly rod and a fly line. While other cast fishing relies on a lure's weight to pull line from the reel during the cast, fly-fishing is a method of casting line rather than lure. The line is heavy enough to send the fly to the target. Artificial flies vary in size, weight, and color, are made by tying hair, fur, feathers, or other materials onto a hook, and are designed to resemble insects, baitfish, or other prey. A "wet fly" is meant to be pulled through the water as though it is an aquatic insect, while a "dry fly" is meant to remain on the surface resembling an insect that fell there. While flies are available commercially, some fly-fishers enjoy tying their own. An experienced fly-fisher selects a fly that is appropriate for the location, weather, and time of year. Izaak Walton, in *The Compleat Angler*, described as many as sixteen flies for each month and how to make them.

The angler holds the rod in one hand and manipulates the line with the other, pulling line out in small increments as the energy in the line, generated from backward and forward motions, increases. The forward cast is most common, where the angler whisks the fly into the air, back over the shoulder then forward, using primarily the forearm. Loops of line should unfurl completely before the angler throws the rod in the opposite direction. When dropping the fly onto the water, the line should land smoothly and the fly should appear natural. If a fish strikes, the angler pulls in line while raising the rod to set the hook in the fish's mouth. Needless to say, mastering these techniques requires a lot of practice. Species sought in fly-fishing are usually trout, including rainbow trout (see color plate C), brook trout, and brown trout, but there is saltwater fly-fishing also. There is concern that as water temperatures rise due to climate change that cold mountain streams may become uninhabitable for trout, which could lead to the extinction of many populations sought by fly-fishers.

Question 3: What is ice fishing?

Answer: Ice fishing involves fishing through a hole cut in the ice of a lake or pond. Shallower ponds and lakes have chain pickerel, northern pike, yellow perch, and sunfish. Deepwater lakes have northern pike, walleye, or lake trout. Brown trout, rainbow trout, and landlocked salmon are often found in deeper lakes, which have the needed cool temperatures in the summertime. However, in the winter, they are caught closer to the surface, just under the ice.

Recreational Fishing in Literature

In literature and film for adults, fishes generally appear as objects, sometimes symbolic, desired by fishermen. *The Old Man and the Sea,* a novella by Ernest Hemingway, published in 1952, was his last major work of fiction. It focuses on Santiago, an aging Cuban fisherman who struggles in an epic battle with a giant marlin far out at sea. Santiago sets out alone, sets his lines, and the big fish takes his bait. Unable to pull it in, he finds the fish pulling his boat. After two days and two nights, both are wounded by the struggle; on the third day, the fish begins to circle the skiff, indicating its tiredness. Santiago uses all his remaining strength to pull it in and stab it with a harpoon. When he heads back to shore, sharks are attracted to the blood and eventually consume most of the marlin. This story was made into a movie, starring Spencer Tracy.

A River Runs Through It, an autobiographic novella by Norman Maclean, concerns a family in early twentieth-century Montana whose views on life are filtered through their passion for fly fishing. The story is presented from the point of view of the older brother who goes on one last fishing trip with his troubled younger brother in an attempt to help him get his life on track. The novella includes detailed descriptions of fly fishing and nature, deals with profound metaphysical questions, and is recognized as an American classic. The 1992 film, directed by Robert Redford, stimulated a rise in fly fishing's popularity.

The book and movie *Jaws* (by Peter Benchley) so demonized sharks that it stimulated a "feeding frenzy" on the part of fishermen. Killing sharks was perceived as a public service until many species became endangered and their ecological importance was realized. Whether this new awareness will save sharks is undetermined.

There are a variety of tools available to cut the hole, including "spud" bars, which work on ice up to about a foot thick, and hand-powered augers. The size of the auger should be relative to the species of fish being sought. Fishing methods include "jigging" and "tip-ups." Jigging involves the use of a jigging rod or hand line and a small jigging spoon or lure which is often baited and darts around in different directions when the angler jerks it up and down. The tip-up is basically a spool on a stick holding a baited line suspended through a hole in the ice. When the bait—usually a minnow—is taken by a fish, the pull on the line releases a signal, such as a red flag. Each state has regulations on ice fishing. For example, in New York, regulations limit anglers to two jigging lines (or hand lines) and five tip-ups. Each tip-up must be marked with the operator's name and address; the operator must be present when the lines are in the water.

Question 4: What is catch-and-release fishing?

Answer: Every angler has probably released or "thrown back" a fish, perhaps because it was below legal size, or it was taken in an area where there were limits on the number of fish each person was allowed to catch, or maybe just because it wasn't wanted. As people have become more conservation-minded, fishery agencies promote catch-and-release fishing in order to provide fishing opportunities while still conserving local fish stocks. Catch-and-release fishing is particularly important in rivers, lakes, and reservoirs where fishermen are numerous and there is a serious risk of depleting fish stocks. In the United States, catch and release was introduced as a management tool to reduce the cost of stocking hatchery-raised trout. Anglers fishing for pleasure rather than for food accepted the idea of releasing the fish. Conservationists advocate it as a way to promote sustainability and avoid overfishing. It is also practiced in areas where the fish have elevated levels of pollutants like mercury or PCBs and should not be eaten. New York City, for example, has a catch-and-release "Big City Fishing" program in Manhattan. The Billfish Foundation runs a tag-and-release program and has added a new "billfish release" certificate for anglers. The information anglers provide goes into a database increasing our knowledge of billfish migration.

Lee Wulff, an influential fisherman and writer, has been credited with popularizing catch-and-release fishing in the 1950s. He pointed out that it may take over a thousand young trout or salmon to produce one or two pair of adults, and the genes of these big, successful survivors are invaluable and should be passed on to the next generation of fish. He urged people who were fishing for food to keep only the smaller fishes, which was the opposite of earlier thinking that advocated releasing the small fishes so they would have the opportunity to grow larger (the basis for legal size limits).

A certain number of fishes die following catch and release due to injury or stress from being out of the water and handled. The mortality rate depends on the species and circumstances. A recent study found that cutting the line is safer than removing the hook when fish are deeply hooked. Almost half the bluegills from which the hook was removed died after ten days, while only 12 percent of those in which the line was cut died. After ten days, most of the fish from the cut-line treatment were able to expel the hook. Some localities require the use of barbless hooks or circle hooks, in which the point is more likely to hook the fish in the mouth rather than be swallowed and cause more damage.

Question 5: How do fishing lures work?

Answer: Basically, the way to catch a fish is to attract its attention with something that looks like or actually is food. The worm or baitfish impaled on a hook is just the beginning. Elaborate molded plastic lures with hooks attached in different shapes, weights, and sizes are designed to tempt the particular fish you want to catch. Simple "spoons," when pulled through the water, flash reflections that resemble an injured or sick fish—tempting a predatory game fish. Poppers burble along the surface, also appearing like an injured fish. Flies may look like actual insects or crustacean prey. There are all sorts of fancy lures, promoted by enticing videos on Web sites, designed to "lure" the angler to part with his or her money in order to catch bigger fishes more easily. For example, there are glow-in-the-dark lures that fishes supposedly prefer even during the day. The glow-in-the-dark pigment is available for anglers to apply to their tackle, float, or tip of the pole. What may be the ultimate lure are soft baits that have been injected with specific pheromones (chemical stimulants) that mimic the pheromones released by a fish to communicate danger. The odor can trigger predators to feed. Then there is the "Secret Weapon Wiggle Rig" that uses an elastic cord, capable of stretching three times its length, between the weight and lure. This setup is said to give the angler greater flexibility in making the lure move realistically in response to slight movements of the rod. These are just a few examples of the wide variety of equipment available to make recreational fishing more exciting and satisfying.

Along with improvements in lures are improvements in boats that now can be equipped with live wells for bait, electronic fish finders, and global positioning systems.

Question 6: What is a party boat?

Answer: A party boat is a commercial sport fishing boat that charges a fee per passenger for regularly scheduled trips. The boat has set departure and return times, holds a set number of passengers, and charges a standard fee. The number of anglers on a trip depends on the size of the boat and the cost of the trip. The lower the cost and the bigger the boat (also the shorter the trip), the more people will be on the boat. General rules on a party boat include following instructions from the crew and captain, being aware of those around you, using patience and courtesy, and being aware of local species and laws. One illegal fish on a boat may result in large fines for the captain and the angler. In addition to fishing tackle, suggested items to bring along include food and beverages (not alcohol), extra clothing, rain gear, sunscreen, sunglasses, an old towel to wipe your hands, shoes that can get wet, a cooler for the fish, tools (like pliers, scissors, and knife, extra line, and a tackle box or bag that can get wet), rope to tie all the gear down, and a waterproof bag to put things in—quite a lot of stuff! Also suggested are cash for tips and snacks, aspirin (or equivalent), and sea-sickness medicine, which, if needed, should be taken before departing the dock. Another option for fishing away from shore is a charter boat that charges a fixed fee for the entire boat and is scheduled around a small set of customers.

Question 7: What are fishing tournaments?

Answer: A shark tournament is a competition for who can catch the biggest shark. Teams of entrants pay a fee and must have a seaworthy boat. Prizes up to thousands of dollars are given to the winners. The captured fish may be given to scientists, to food pantries, or may be just discarded. These tournaments became popular in the 1970s after the publication of Peter Benchley's book *Jaws* and the subsequent movie, but now that so many sharks are threatened and endangered, tournaments have become a focus of protests by conservationists and humane organizations. In the last five years over a half million sharks were harvested annually by the sport-fishing community in the United States alone. Many of these were breeding-age animals of vulnerable or endangered species. Research has shown that removal of adult sharks from the population is occurring at such a rate that many species stand no chance of survival. Conservationists are concerned about the long-term survival of the ocean's top predators. Scientists with the IUCN (International Union for the Conservation of Nature, formerly the World Conservation Union) have found that eleven shark species are on the high-risk list, while five others are showing signs of decline. Some of the more threatened species are the very ones that are most prized in shark tournaments. Over the past thirty years, the IUCN reported, mako and thresher sharks have lost about 75 percent of their population and are classified as vulnerable and threatened with global extinction.

A new effort, called Shark-Free Marinas aims to reduce worldwide shark mortality. This nonprofit company intends to prevent the deaths of millions of sharks by working with game-fishing societies, tackle manufacturers, competition sponsors, and marinas to change attitudes and policies and promote catch-and-release fishing. The initiative works by prohibiting the landing of any shark at a participating marina. In 2009, organizers of a Fort Myers Beach shark tournament announced they will no longer allow sharks to be killed and held a catch-and-release tournament.

Tournament fishing for catfish is a growing sport, and most tournaments have a catch-and-release policy to ensure that big fish remain in the waters. This is ecologically beneficial, as the largest female fish are disproportionately responsible for producing most of the eggs for the next generation. In Virginia, each participant is allowed to keep only one blue catfish over 32 inches long in a day of fishing. Three species are prized: flatheads, *Pylodictis olivaris;* blue catfish, *Ictalurus furcatus;* and channel cats, *Ictalurus punctatus.* Tournament winners are usually partners who cooperate in running the boat and doing the actual fishing, and they share a cash prize for the biggest catch, which may be the combined weight of no more than five fish. Bass tournaments are also very popular.

So what are some tips to help you catch the big ones? James W. McKenzie, director of the Catfish Anglers Association, said that if he had to suggest one thing to improve success it would be bait. Large catfish generally feed on live fish, so to catch them you need large fresh bait. Equipment, location, weather, and experience are also important. The equipment should be appropriate for the fish you are after—don't bring a fishing rod suitable for a 5-pound fish when you are after a 30- to 40-pound fish. Location, just like in real estate, is very important. Catfish, like other fishes, love structures like rocks, branches, or pilings and will stay around a structure to feed and rest. The habitat

Fishing as Therapy

A project called Project Healing Waters Fly Fishing serves military personnel who have been injured in battle. Its goal is to assist their physical and emotional recovery by learning fly fishing and going on outings. Veterans can leave their hospitals and be out in nature in a safe environment to improve dexterity, depth perception, motor skills, strength, and balance. Many of them develop not only a life-long love of fly fishing but of conservation also. A similar program called Casting for Recovery runs two-week-long fly-fishing retreats for breast cancer patients. The women learn the techniques of casting, knot-tying, and using equipment. The casting process is particularly beneficial for the arm muscles, and increased blood flow aids in healing. The patients also experience emotional and social benefits as they spend time on the water with supportive people.

does not have to be large—it can be a small rock that breaks current and allows the fish to save energy by sitting behind it. In tournament fishing you are at the mercy of the weather, and knowing how to fish in that weather is helpful. You can find photos of tournament winners (both the fish and the anglers) and more information and suggestions on the Catfish Anglers Association Web site.

Question 8: What is the most fish anybody has ever caught in one sitting?

Answer: As part of a fund-raiser for *Fishing for Life,* a group that raises money for urban youth ministries in Minneapolis and St. Paul, a contest was held to see who could catch the most fishes in twenty-four hours. The winner was Jeff Kolodzinski, who set a new world record, according to the newspaper report, by catching 1,628 fishes from noon on August 22, 2008, to noon on August 23 on the shores of Lake Minnetonka. He reportedly used eight thousand maggots as bait and caught mostly bluegills and pumpkinseed sunfish. During the first two hours he caught a fish every twenty-two seconds. On average, he caught one fish every fifty-three seconds, although things slowed down during the night. He admitted that he nodded off, but fish biting on his line would awaken him. The article does not specify whether he was releasing the fishes and possibly catching many of the same fish repeatedly.

Question 9: Is recreational fishing regulated?

Answer: Freshwater fishing is regulated in each state, while marine fishing is regulated in coastal states. Freshwater anglers over a certain age need to purchase fishing licenses, but saltwater anglers don't necessarily need them. Check with your state to learn the regulations before going fishing. Regulations apply to particular species and include season, places where fishing is allowed, minimum size, and numbers (bag limit) that can be caught. Some states may also have a maximum size for certain species, as Texas does for red drum to conserve breeding stock. Regulations may also apply to types of gear that may be used.

However, while commercial fishermen must submit extensive data about their catch, there is no requirement for recreational fishers to report their catch. States rely instead on estimates from telephone surveys and random site visits. The lack of documentation leaves great uncertainly about the true impacts of recreational fishing, and some states are trying to improve their data. North Carolina has started a pilot project using text messaging, where recreational fishers send catch information to a computer database. In Massachusetts anglers enter information into a Web-based reporting system to allow the Division of Marine Fisheries to analyze catch data more accurately.

NINE

Commercial Fishing

Question 1: How does commercial fishing differ from recreational fishing?

Answer: Commercial fishing is the activity of capturing seafood for commercial profit. Commercial fishermen harvest a wide variety of animals, including finfishes like tuna, cod, and salmon, but also invertebrates like shrimp, lobsters, clams, and crabs. Large-scale commercial fishing is also called industrial fishing. A fishery, according to the FAO (Food and Agricultural Organization of the United Nations), is defined as the people involved, type of fishes, geographical area, and method of fishing. The term is often applied to the fishes in a region and to those who are fishing for similar species with similar gear. Commercial fishing is a dangerous profession, with an average annual fatality rate of 115 deaths per 100,000 fishermen in the United States due to falling overboard, capsizing, or sinking. Commercial fishing vessels are required to carry life rafts, radios, and immersion suits to reduce fatalities.

Question 2: How did the fishing industry develop?

Answer: Fishing as a commercial enterprise began in eleventh-century Europe, where fishermen used nets, spears, traps, weirs (low dams across rivers), and hook and line to catch salmon, shad, whitefish, and sturgeon. Over time, larger boats were built that could venture further out to sea and harvest more fishes.

As the human population grew, there were larger markets and a need for more fishes, so larger nets and gear were produced. In the nineteenth century, lines with one hook were replaced by long lines with hundreds or thousands of hooks that could extend for miles. Gill nets were developed that did not need bait, as fishes trying to move through it got caught by their gills. Traps were enlarged and combined into complicated systems with many traps. New boats and equipment in the 1500s made it possible to venture from coastal to deep-sea fishing.

A major advance was the development of nets that were dragged behind boats. The beam trawl, invented in fourteenth-century England, was a bag net held open by a long wooden or steel beam, towed by sailing vessels and later by steam-powered vessels. Its initial use prompted opposition, for fishermen could see the damage it caused to their favorite areas. Despite attempts to ban trawls, the method spread, for it was very efficient. Over generations fishermen began to accept them. Trawls were developed in which the bottom opening was maintained with a chain with rollers beneath it, the top was held up by floats, and the sides of the net were kept wide open with wooden "doors" on each side. This otter trawl, named after the "otter boards" that held it open, became standard equipment.

By the late 1800s, sailing boats were replaced by steamships, and then motorized vessels early in the twentieth century. Large ships had deck space for onboard fish processing—the forerunner of "factory ships." The ability to freeze fish on board was a major advance in the mid-twentieth century. Echo sounders and sonar allowed crews to locate and catch schools of fish more efficiently. While mechanical tools like the winch and pulley were introduced in the early twentieth century, during the 1970s and 1980s ships became more automated.

As gear improved, new fishing techniques were introduced. Larger nets could be manufactured more efficiently, and traditional materials were replaced with synthetic materials that lasted longer. Recently, the rollers on the bottom of trawls have been replaced by "rockhoppers" (fig. 23), large disks that hop up when they hit a rock, making it possible to drag the net over a

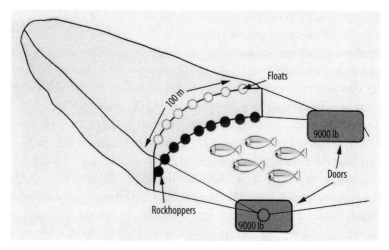

Figure 23. Otter trawl with rockhoppers. *(Courtesy of NOAA.)*

rough, rocky bottom without getting torn up. Trawls with openings large enough to accommodate a dozen 747 airplanes are
in use. These technological advances in boat and gear size and
efficiency made it easier to catch more fish faster, so most of the
world's commercial catches increased until the 1980s, when this
efficiency led to depletion of many populations. As fish populations declined, vessels went to new areas and increased their
fishing effort, thus depleting the populations even more. The
industry is dominated by large boat, industrialized fishing, and
small-boat fishermen who have had much less impact on fish
populations have become an endangered species, along with
the fish.

Question 3: What is the importance of cod to the history of New England?

Answer: Cod, *Gadus morhua* (see fig. 19, in chapter 5), has been
an important economic commodity since the Viking period
(around AD 800). Prior to refrigeration, having a fish product
that would not rot and could be dried, stored, and still be tasty
was important. Cod has very little fat and, when dried, is almost

80 percent protein. Norwegians used dried cod during their travels and a market developed in southern Europe. Since the introduction of salt, dried salted cod has become more important. The Basques also played an important role in the cod trade and are claimed to have found the Canadian fishing banks before Columbus. The New England colonies developed and prospered in part due to the vast amount of cod; many New England cities are near cod fishing grounds. In Massachusetts and Newfoundland, cod became the major commodity, responsible for turning New England into a commercial power (fig. 24). After Britain began taxing the American colonists, the cod trade grew through trading with French Caribbean colonies for molasses. Residents of New England included a "codfish aristocracy," who traced their fortunes back to the seventeenth-century cod fisheries. Codfish images were found on family crests and state seals. At the end of the Revolutionary War, negotiations were held with England that, after much wrangling, ultimately permitted Massachusetts

Figure 24. Cod fishery. *(Courtesy of NOAA.)*

to continue fishing in the Grand Banks, off the coast of Canada, which remained loyal to Britain.

In the late twentieth century, cod off the coast of Europe and America crashed. Catches initially grew, before they plummeted, because the efficiency of modern trawlers made it possible to locate and catch the remaining fish and remove them. Estimates of stock sizes were based on the fish brought to market, not on the catch, much of which was dumped back into the ocean because of quotas and size limits. In 2000, cod was placed on the list of endangered species by the World Wildlife Fund (WWF), since global catches had dropped by 70 percent over the past thirty years. The necessity of restricting catches to allow populations to recover is opposed by the fishing industry; politicians are reluctant to approve measures that will result in job losses.

Question 4: Which species are caught commercially?

Answer: Hundreds of different species are caught commercially around the world. Among the major groups are Gadidae (cods and hakes), Thunnidae (tunas and mackerels), Salmonidae (salmon and trout), Pleuronectidae (flatfishes), Clupeidae (herrings and anchovies), Percidae (perch-like fishes, snappers, and groupers), catfish, and sharks. The top species caught in the United States in recent years are pollack, *Pollachius;* walleye, *Sander vitreus,* the only freshwater species in the group; Atlantic menhaden, *Brevoortia tyrannus;* Pacific cod, *Gadus macrocephalus;* pink salmon, *Oncorhynchus gorbuscha;* Pacific sardine, *Sardinops sagax;* sockeye salmon, *Oncorhynchus nerka;* yellowfin sole, *Limanda aspera;* and Pacific hake, *Merluccius productus.*

The species that are caught the most are ones that have traditionally been the most abundant, easy to catch, and tasty, although there are fads of particular species that come and go and regional differences in what fishes are considered the best to eat. It is unfortunate that some of the most popular species continue to be fished even after their populations have plummeted.

Question 5: What kind of gear is used to catch them?

Answer: Most commercial fishing is done with nets. Trawls are towed along the bottom behind a boat (see fig. 23). Purse seines are walls of netting that surround and enclose a school of fish—a rope passes through rings on the bottom and, when pulled, draws the rings close to one another, like a drawstring on a purse, preventing the fish from swimming down to escape. Gill nets hang in the water, and when fish try to swim through the openings, they cannot squeeze through swimming forward, and when they try to back out, they get caught by their gill covers. Another category is lines with hooks. Longlines deployed from fishing vessels may be many miles long and have thousands of hooks. Pelagic longlines hang near the surface to catch fish like tuna and swordfish, while benthic longlines lie on the sea floor for bottom-dwellers (groundfish) like halibut or cod. Then there are traps. They are placed in a fixed location and might consist of a frame of thick steel wire with netting stretched around it. The mesh wraps around the frame and then tapers into the inside with funnel-shaped openings. Fish follow the mesh as it curves into the funnel, swim in through the funnel-shaped opening, and find themselves inside the chamber—they generally cannot get out. More complicated trap nets, pound nets (fig. 25), fyke nets, or weirs may consist of walls of netting in a pattern that directs fish into the mouth through funnel-like openings.

Question 6: Which species are overfished and what effect does overfishing have on the species?

Answer: Overfishing occurs when fish are taken faster than the stock can replenish itself or if the stock size is too small to be sustainable at current fishing levels. People expressed concern about overfishing in the early twentieth century, but through the 1940s fish yields increased because fleets fished in new areas

THE FISHERIES OF THE GREAT LAKES.
"Lifting the pot" at Kelley's pound-net, Lake Erie. (Sect. v, vol. i, p. 760.)
Drawing by H. W. Elliott.

Figure 25. Pound net. *(Courtesy of NOAA.)*

or for different species. The opening of new fishing grounds hid the population declines. Between 1950 and 1994, the total catch increased by 400 percent because larger fishing fleets had greater technological power to catch fish. In 1997 and 1998, the global capture peaked at 93 million tons, then declined due to catching more than what the ocean can produce. Ten years ago, the U.S. Department of Commerce declared 98 species overfished. In 2008, the European Commission declared 88 percent of fisheries stocks of the European Union (EU) overfished. Scientists estimate that 90 percent of large fish—like tuna and swordfish—have been removed from the oceans. Stocks of valuable, large species (such as bluefin, *Thunnus thynnus;* and bigeye tuna, *Thunnus obesus*) have been severely depleted and are a serious concern. Bluefin tuna (fig. 26) can sell for over one hundred thousand dollars apiece. Their populations have declined so greatly that there was a move to ban fishing under CITES (Convention on International Trade in Endangered Species). Some of the depletion is due to illegal fishing, particularly in Asia and Africa, where countries lack the resources to patrol their waters.

Figure 26. Bluefin tuna, *Thynnus thynnus*, caught in net. *(Courtesy of NOAA.)*

Overfished species include Atlantic cod, *Gadus morhua* (see fig. 19, in chapter 5); tunas; European hake, *Merluccius merluccius;* Atlantic halibut, *Hippoglossus hippoglossus;* Dover sole, *Microstomus pacificus;* salmon; monkfish, *Lophius;* marlins; swordfish, *Xiphias gladius;* and sharks. Deep-sea fishes like orange roughy, *Hoplostethus atlanticus;* Patagonian toothfish, *Dissostichus eleginoides* (aka "Chilean sea bass"); and sablefish, *Anoplopoma fimbria* are at risk. They grow slowly, have low reproductive rates, and may not mature for thirty to forty years. A fillet of orange roughy may be over fifty years old. Most deep-sea fishes are in international waters where there are no legal protections; they are caught by deep trawlers near seamounts, where they congregate because of food. Flash freezing allows trawlers to work for days at a time. Scientists are also concerned that smaller species, like herrings, anchovies, and sardines, a crucial part of the ocean's food chain, are also under siege. They are eaten not only by fish, but also by many people throughout the world.

On the other hand, there is some good news. Most U.S. fish stocks are not overfished. In May 2009, the National Marine Fisheries Service reported that four stocks—Atlantic bluefish, Gulf

of Mexico king mackerel, and two stocks of Atlantic monkfish—
have been rebuilt to allow for continued sustainable fishing.
This is the largest number of stocks to be declared rebuilt in
one year since the fisheries service declared the first stock successfully rebuilt in 2001. In the summer of 2009, a team of scientists led by Boris Worm of Dalhousie University and Ray Hilborn of the University of Washington (who previously had major
disagreements) assessed marine fisheries around the world and
concluded that steps taken to curb overfishing were having
positive effects in many ecosystems including the United States,
Iceland, and Kenya. Catch restrictions, gear modification, and
closed areas are helping to rebuild overexploited stocks. Nevertheless, about two-thirds of the stocks were still in poor condition, and lower catch rates are needed to reverse the collapse.
Restrictions have long-term benefits but come with short-term
costs to fishermen.

The increasing demand for fish is due to population growth
and increasing prosperity in the developing world. According to the United Nations Food and Agriculture Organization
(FAO), more than 75 percent of the world's fish stocks are fully
exploited, over-exploited, or recovering from past depletion.
Tens of thousands of bluefin tuna (see fig. 26) were formerly
caught in the North Sea annually, but now there are none. Once
there were millions of skates around the United Kingdom; now
most are gone. As fishing pressure increases around coral reefs,
larger species disappear, followed by smaller species. One can
see enormous differences over time; many fishes once abundant
are gone from Caribbean reefs. The growing appetite for live
reef fish across Southeast Asia and China is devastating populations in the Coral Triangle, a region with the world's richest
ocean diversity. Spawning of reef fish in this area (containing
three-fourths of all known coral species in the world) has declined by 80 percent, according to a report by Yvonne Sadovy
at the University of Hong Kong. Overfishing of spawning aggregations, when fish gather in great numbers to reproduce, may
be the culprit. Dr. Sadovy suggests that spawning aggregations

be considered "protected events" rather than simply times when fish are easy to catch.

In addition to reducing the population, the size of individual fish gets smaller. Archaeological records show that average size of cod (see fig. 19, in chapter 5) in the Gulf of Maine remained similar over the past five thousand years until rapid declines in the late twentieth century. One hundred years ago, cod measuring 5 feet or more were common, while today the biggest specimens are under 2 feet; and 18- to 20-foot sharks, common in 1700, are unusual today. In 2009, when a trawler caught an 8-foot-long halibut, it made the news, while fish that size used to be common. Each generation views as "natural" the environment they remember from their youth, the phenomenon called "shifting baselines." Loren McClenachan of the University of California, San Diego, used archival photographs of trophy fish (fig. 27) caught in Florida to document the dramatic decline of their size from the 1950s to the present. Most of the big fish disappeared. There was also a shift in trophy-winning species—the huge groupers caught in earlier years disappeared, and smaller snappers became trophy fish.

Fishing has evolutionary consequences. It acts as "selection pressure" against the large fish that are targeted. After decades of heavy fishing, Atlantic cod (and other species) evolved to reproduce at younger ages and smaller sizes, allowing them to leave offspring before they are caught. However, fishes that reproduce earlier produce far fewer eggs than those that grow another year (see chapter 5: "Fish Reproduction"). Commercial fishing also appears to be wiping out the cod that swim at shallow depths and that are genetically different from those in deeper water. Icelandic cod stocks are presently in far better shape than the collapsed fisheries in the western Atlantic. Einar Árnason of the University of Iceland studied their population genetics to find out how genes changed between 1994 and 2003 and found that the gene associated with shallow water is rapidly decreasing, which is logical since the fishery targets fish in fairly shallow waters. They also found that Icelandic cod are

Figure 27. Trophy fish caught on charter boats in Key West: A, 1957; B, early 1980s; and C, 2007. (*Photos from Loren McClenachan.*)

becoming sexually mature while still small and young, the same thing that happened in western Atlantic cod just before that fishery crashed.

However, this evolution is reversible. David Conover of Stony Brook University subjected laboratory populations of silversides,

Menidia menidia, to size-selective fishing (removal of the largest 90 percent of each generation) for five generations, resulting in fish that grew to a smaller maximum size. Then the size-selective harvesting was stopped for another five generations. The populations that had evolved smaller body size during the first part of the study increased body size when the simulated fishing stopped.

Question 7: How does large-scale fishing affect the ecosystem?

Answer: Imbalances are appearing in marine ecosystems. Boris Worm of Dalhousie University found that as large fish disappear, the fishing industry shifts its focus down the food chain, taking more small plankton-eating fish like sardine and anchovy. This "fishing down the food web" is of concern since smaller fish are required by larger predatory fish as well as birds and marine mammals that eat them. They also help to maintain balance in the species below them in the food chain, like jellyfish. Blooms of jellyfish have appeared in many overfished waters. The Benguela current off the coast of Namibia in Africa once supported large populations of sardines and anchovies, but now has been taken over by jellyfish with a biomass over three times that of mackerel, hake, sardine, and anchovies combined. Since jellyfish eat fish eggs and larvae and compete with young fish for food, the shift to a jellyfish-dominated system may be irreversible. With overfishing causing ecological shifts, recovery is not always possible even if fishing is stopped. Overfishing is not only removing marine life, it is obliterating habitats and destroying the resilience of ecosystems.

Removing fish from an ecosystem may have consequences on organisms lower on the food web. A decrease in phytoplankton-eating fish can result in phytoplankton blooms, which can lead to hypoxia (low oxygen) in the water. The collapse of cod, haddock, hake, pollock, plaice, and flounder fisheries off Nova Scotia coincided with an increase in phytoplankton blooms, traditionally attributed to nutrient pollution, and increases in sea urchins, crabs, and lobsters. Squid are thriving throughout the oceans, due largely to removal of their predators. In addition to these "top-down" effects on prey species, other species may be dependent on fish diversity. For example, the appropriate balance of herbivorous fishes on coral reefs keeps algae from smothering coral reefs.

Question 8: How can commercial fishing techniques affect the environment?

Answer: Some kinds of gear damage the environment. Trawling involves towing a net along the bottom (see fig. 23), not only catching any fish in the way but also mowing down the structure on the bottom, destroying habitat. The bottom can change from a biologically rich, productive habitat to a vast expanse of bare sand and mud. This is like clear-cutting a forest, but not visible to us. The bottom in the Gulf of Maine was once thick with mussels, crabs, anemones, and tubeworms, but after trawling it appeared bulldozed. There is particular concern for deep-sea mounts with corals that are destroyed by trawls with rockhoppers. Sea mounts draw currents up from the sea bed and are highly productive regions supporting large numbers of fishes that had previously been inaccessible to fisheries. Now these long-lived, slow-growing, deepwater species are being targeted and their habitat mowed down. Fortunately, these destructive fishing practices are being restricted, and miles of sea floor are now protected from bottom trawling.

In addition to the species that are being sought, many other fishes are caught accidentally and are called "bycatch." Bycatch is generally discarded—dead—back into the water; it can include juveniles of many species that may be commercially important when grown, but killing them before they can reproduce depletes the population. Dolphins and sea turtles, which are protected, and sharks with declining populations can also be bycatch. Some fisheries catch more bycatch than the targeted species. Longline fishing, with thousands of hooks, produces a lot of bycatch, including marine birds that are attracted to the bait, caught on the hooks, and drowned. Traps and gillnets may be lost, but they continue fishing, continuing to catch and kill for years. This so-called ghost fishing is another major waste. It is estimated that somewhere around 4,000 miles worth of lost and discarded gillnets are on the bottom in the North Atlantic. (As part of the "stimulus package" to improve the U.S. economy, fishermen in Puget Sound are being paid to haul up lost nets from the bottom. They expect to pull up about three thousand nets.)

Drifting fish aggregating devices (FADs) are used in purse seine fisheries. Over 1 million tons of tuna and over 100,000 tons of bycatch are caught near FADs each year. FADs may contain sonar equipment and GPS (global positioning system). A vessel can remotely contact a FAD via satellite and check sonar readings to determine the size of a fish school. The widespread use of FADs has shifted the pattern of exploitation of tunas over the past twenty years, with smaller tunas being caught now than when purse seiners targeted mostly free-swimming schools.

Gear can be designed that reduces bycatch. Turtle excluder devices (TEDs) are now required on shrimp nets that previously caught large numbers of turtles. Longline fisheries can reduce turtle bycatch by replacing J-shaped hooks with circle hooks that are less likely to catch turtles, and if caught, turtles are less likely to swallow those hooks. Fishing boats can catch fewer birds by tying red plastic streamers to the vessels. There is the International Smart Gear Competition that awards prizes to those who come up with practical, cost-effective, and innovative designs to reduce bycatch.

Question 9: What regulations govern the fishing industry?

Answer: Regulations are set in the United States by regional fishery management councils, consisting of members of the fishing industry plus scientists and government representatives. They make recommendations to the National Marine Fisheries Service, part of the National Oceanic and Atmospheric Administration. In areas outside national control (200 nautical miles) international agreements are required. The need for agreements on fishing and other maritime issues led to conferences on the Law of the Sea and a treaty known as the United Nations Convention on the Law of the Sea. Some areas need additional intergovernmental coordination. For example, in the Mediterranean Sea and other narrow bodies of water, zones of 200 nautical miles are irrelevant and international agreements are needed. Also, oceanic species like tuna come under no one's jurisdiction, so there are international groups such as the International Commission for the Conservation of Atlantic Tuna responsible for their management. Since tuna have declined by over 90 percent and the remaining ones continue to be fished out of the oceans, the effectiveness of these commissions is doubtful.

Fishing operations can be regulated by vessel licensing and restrictions on catching techniques. Direct methods of regulation include limited entry, catch quotas, and prohibition of certain techniques. Traditional fishery management involves setting total allowable catch levels based on population assessments by fishery scientists, which are very uncertain. Fishery scientists recommend a "maximum sustained yield," the maximum catch that should be allowed. Because of uncertainty in population estimates, management councils tend to err on the side of allowing more, rather than less, fishing, setting quotas higher than scientists recommend. Setting quotas causes fishermen to compete with one another to catch the most fish sooner, before the quota is reached. After the quota is reached, fishermen target a new species, but still catch the first species as bycatch that now has to be discarded dead, since they may not be landed. Often the quota cannot even be reached because it was set too high and there are not enough fish left. Thus, a quota may not protect the fish at all.

Traditional management sets minimal sizes, with the aim of reducing the number of big fish, leaving more resources for younger fish. Most marine fish produce huge numbers of eggs, and the assumption has been that younger spawners will produce plenty of viable larvae. However, in many species, large, old females are far more important than younger fish since their larvae are more numerous, grow faster, and are more likely to survive than offspring of younger fish.

Camilo Mora of Dalhousie University evaluated management effectiveness of fisheries around the world and found that only 7 percent of all countries carry out rigorous scientific assessment, 1.4 percent also have a participatory transparent process to convert scientific findings into policy, and less than 1 percent also ensure compliance with regulations. No country was free of the effects of excess fishing capacity, subsidies, or access to foreign fishing. Wealthier countries have better science and enforcement but have excessive subsidies and fishing capacity due to modernization of their fleets. Poorer countries lack robust science and enforcement; and, although they have less fishing capacity, they sell fishing rights to nations that do. In one-third of low-income countries, most of the fishing is done by foreign fleets.

In a rare use of the precautionary principle (if an action or policy might cause severe or irreversible harm to the public or the environment, in the absence of a scientific consensus, err on the side of safety), the North Pacific Fishery Management Council voted in 2009 to prohibit commercial fishing on nearly 200,000 square miles of U.S. waters in the Arctic—waters off Alaska north of the Bering Strait and east to the Canadian border. No large-scale fishing operations currently exist, although subsistence fishing by indigenous people would be allowed. Conservation groups and the fishing industry backed the decision, based on lack of data. In the summer of 2009, the panel that regulates fishing in the Gulf of Mexico voted to limit the use of longlines for grouper because the lines can snag and drown threatened loggerhead turtles.

Question 10: How does overfishing affect the economy?

Answer: The World Bank and the Food and Agricultural Organization (FAO) concluded that the depletion of fish stocks constitutes a loss of a nation's wealth; the inefficient state of global fisheries causes losses of fifty billion dollars annually.

Because of overcapacity, excess fleets compete with each other for limited fish resources, causing inefficiency. With less fish to catch, the cost of fishing is greater. They recommended removal of the open-access condition of marine fisheries and an end to subsidies that create excess fishing capacity and effort.

The economic situation is critical for small-scale fishing in developing countries where fisheries provide an important source of livelihood for 1.5 billion people and where fishes provide over 20 percent of animal protein consumed, compared to 8 percent in industrialized countries. Small-scale fisheries provide a large percentage of the global catch, and declining fisheries cause increased poverty, according to the U.S. Agency for International Development.

Fisheries and Pirates

Pirates off the coast of Somalia received a lot of attention in 2009 when they attacked commercial ships and held crews for ransom. But the connection of pirates to fisheries is not as well known. Since the civil war that brought down Somalia's last functional government in 1991, the country's coastline has been pillaged by foreign fishing vessels. Without a coast guard, Somali waters have become the site of an international fishing free-for-all, with fleets from around the world, including South Korea, Japan, and Spain, operating along the coast, often illegally and without licenses. Somali fishermen, whose industry has always been small-scale, lack the advanced boats and technologies of the foreign competitors and complain that foreign fishermen are stealing their fishes and shooting at them. Pirate gangs, including some former fishermen, emerged initially in the 1990s to protect against foreign trawlers. Piracy has since become a lucrative industry, and most of the current pirates are probably not former fishermen but others seeking to make a fortune. The environmental costs of mismanaged fisheries, therefore, can go much further than the damage to fishes and their ecosystems.

Question 11: What are "fish wars?"

Answer: The oceans belong to no one, and when fish stocks diminish, the potential for international confrontation increases. Throughout the world, incidents happen where one nation's fishing fleets go into fishing areas considered under the jurisdiction of another country. Conflicts between England and Iceland over cod started in the 1500s. Steam power in the late nineteenth century allowed exploitation of new areas, and large cod catches around Iceland stimulated British ships to fish there. In 1893, the Danish government (that governed Iceland) claimed a fishing limit of 13 nautical miles, but British trawlers continued to fish inside Icelandic waters. Danish gunboats escorted a number of vessels to port, fined them, and confiscated their catch. An 1896 agreement allowed British vessels to use Icelandic ports for shelter, but they were not allowed to fish in certain areas. In 1899, a Danish gunboat tried to arrest a trawler for fishing inside the limits. The trawler refused to stop and was fired on. The captain was arrested, convicted on several counts, and jailed. With many British trawlers being charged and fined by Danish gunboats for fishing illegally (within a limit that they did not recognize), the press began to inquire why this action against Britain was allowed without intervention by the Royal Navy. The issue was left unresolved when World War I effectively ended the dispute.

Three more cod wars took place in the twentieth century as Iceland progressively increased its exclusion zone, and the British did not recognize the expansion and declared that their trawlers would fish under protection from their warships, and Icelandic vessels collided with British vessels until an international body intervened. The Third Cod War (November 1975–June 1976) occurred after Iceland increased the zone to 200 nautical miles. When Iceland threatened closure of a NATO base, the British agreed to stay outside the exclusion zone. After Iceland's 200-mile zone was recognized in 1976, many other countries declared their own 200-mile zones.

Use of Fish Products for Things Other Than Food

Fish skins can be used to make leather. The skins are purchased from commercial fisheries or canneries and treated with chemicals to remove the oils. Leather is made from skins of carp, Pacific salmon, bass, sturgeon, shark, catfish, tilapia, salmon, and Nile perch. Through chemical and mechanical processes, the skins are churned, soaked, fleshed, and vacuum-dried. Manufacturers can produce many different textures of fish leather, including suede, silk, glazed, pearlized, and high-grain finishes all from the same species. The suede can be water-proofed; the glazed is scratch/stain resistant.

Nearly one-third of the world's wild-caught fish are not eaten by people but turned into fish meal and fed to farmed fish, cattle, and pigs. Aquaculture alone consumes an estimated 53 percent of the world's fish meal and 87 percent of its fish oil. Oil and fishmeal are commonly derived from menhaden, *Brevoortia tyrannus,* also called mossbunkers (see fig. 12, in chapter 4). Atlantic menhaden are phytoplankton feeders and are important prey for striped bass, bluefish, mackerel, tuna, drum, and sharks, as well as for birds and marine mammals.

The important commercial fishery known as a "reduction fishery" is responsible for the extracting the oils (high in omega-3 fatty acids) for human consumption, aquaculture, pet foods, and livestock feeds as well as for producing water-resistant paints, skin treatments, and cosmetics. The fishery began in New England in the early 1800s and spread south after the Civil War. There are two companies that harvest menhaden in the United States: Omega Protein Corporation, in Houston, Texas, and Daybrook Fisheries, in Empire, Louisiana. There is also a bait fishery that harvests menhaden for commercial and recreational fishermen.

My colleague H. Bruce Franklin's book, *The Most Important Fish in the Sea,* is an interdisciplinary study of the role of menhaden in American environmental, economic, social, political, and cultural history. Native Americans taught the

Use of Fish Products for Things Other Than Food

Pilgrims to plant it as fertilizer with their corn, making large-scale agriculture possible in the rocky soils of New England. During the Industrial Revolution, menhaden oil was used as a lubricant in factories, and dozens of factories processed the fish into oil and fertilizer. For over a century, menhaden have been the nation's largest fishery. In 2006, 376 million pounds of menhaden were caught in Maryland and Virginia waters (Chesapeake Bay and Atlantic Ocean), valued at approximately $22.8 million. There is increasing concern from conservationists that the Chesapeake Bay's menhaden population is declining significantly and leading to reduced health of striped bass and other predators.

Another international dispute took place in 1995 when Canadian officials boarded a Spanish trawler, the *Estai*, just outside Canada's 200-mile zone and arrested its crew. The trawler had been using nets with openings that were far smaller than international regulations allowed for turbot fishing. The settlement involved smaller quotas for turbot and tougher inspection and enforcement of fishing on the high seas.

Question 12: Is aquaculture a possible substitute for commercial fishing?

Answer: Hunting and gathering play a small role in obtaining food from the land since most food is grown (agriculture). But most seafood is still obtained by fishing. Aquaculture is the aquatic equivalent of agriculture. As fisheries have declined due to overexploitation, the farming of fishes, invertebrates, and seaweed has become more and more common. Fishes comprise about two-thirds of all cultured animals, and the industry is growing rapidly, about 9 percent per year. Aquaculture now accounts for about half of the fish consumed globally and provides jobs for many people in developing countries. The yield from freshwater aquaculture can be as high as from agriculture in the same space (up to 100 metric tons per hectare—2.4 acres—per year). Aquaculture reduces the pressure on overexploited wild fish populations. Most salmon in restaurants and fish markets is now cultured, not wild, and some local stocks may be recovering.

Aquaculture operations may use existing ponds, build large tanks, or use cages placed in natural bodies of water. Some species are raised through their entire life cycle, while juveniles of other species are caught in the wild and then raised. Finding the proper foods for larvae and juveniles can be difficult. Some low-intensity aquaculture operations allow the fish to eat natural food in the environment, while more intensive operations provide food. The density of cultured fish frequently requires that the water be aerated and/or constantly flowing through the system. Marine systems are often cage cultures that require intensive feeding but do not require land conversion, as is often the case for freshwater systems.

Aquaculture, relatively new in North America, has been practiced for centuries in Asian countries, which still dominate the industry, providing about three-quarters of the world's harvest. Of the fishes grown commercially, most are carps in the family Cyprinidae grown in artificial ponds. Centuries of trial and error have resulted in considerable expertise about the best way to feed and raise the cultured species. Wastes may be used to fertilize the water to stimulate growth of algae, which the herbivorous fish then eat. Other cultured species include Nile tilapia, *Oreochromis niloticus*, in fresh water, and Atlantic salmon, *Salmo salar*, in salt water. Channel catfish, *Ictalurus punctatus*, farms in the Mississippi Delta use ponds and are the major U.S. farmed fish. In April 2009, the Clean Seas Company of Australia announced a breakthrough in captive breeding and raising of tuna, as the first step in developing aquaculture. Similar advances are being made in Japan for bluefin tuna. Since tuna grow very large and must continually swim to obtain enough oxygen, they require much larger pens than other fish. Some culture operations raise ornamentals for the tropical fish pet trade. The development of culture techniques for seahorses is a major breakthrough, as they are rapidly disappearing in the wild due to increasing demands from Eastern medicine and from the aquarium trade.

There have long been attempts to increase wild populations by rearing young fish in hatcheries and releasing them, but this may have unintended negative effects on the population. Offspring of steelhead trout raised in hatcheries for release to the wild were less successful at reproducing. A fish born in the wild as the offspring of two hatchery-reared steelhead averaged only one-third the reproductive fitness of a fish with two wild parents, according to a study by Michael Bluin of Oregon State University. Fishes that do well in the safe hatchery environment are different from those that do well in the predatory real-world environment. Captive breeding and adding hatchery fish to wild populations may ultimately be hurting efforts to sustain wild fish.

Aquaculture can have negative environmental effects. Intense fish culture produces a lot of waste which, unlike most human waste, does not go to a sewage treatment plant and may cause pollution problems. Salmon cage culture is of particular concern as "dead zones" have been found below cages in areas without strong currents to dissipate the wastes. Pollution can harm not only native species, but the aquaculture operation itself, which will suffer from poor water quality. Water quality is also affected by antibiotics, anti-fouling paints, and anti-parasite medications used in salmon aquaculture.

The "crop" may escape from confinement and may become invasive. This has happened repeatedly with tilapia (a common name for nearly one hundred species of cichlids belonging to three genera), which have displaced native species on several continents. The only way to really prevent invasion is to not culture species outside their native range. Furthermore, domesticated varieties may be genetically modified for fast growth, and they could displace natives of their own species if they escape. Another concern is movement of disease or parasites from cultured to wild fish in the vicinity, as has occurred with "fish lice" on cultured salmon (see color plate G) moving onto migrating wild salmon as they pass by.

Finally, there is the ecological concern that in order to raise carnivores like salmon, huge quantities of smaller fishes are used for fish meal and oil. As aquaculture has grown, its share of fish meal and fish oil consumption has more than doubled over the past decade to over two-thirds and almost 90 percent, respectively. The fishes used to make fish meal are themselves limited, and their overexploitation will ultimately result in less food for the wild fishes that ordinarily eat them. Using fish meal to feed farmed fish is also inefficient. About three pounds of forage fish go to produce one pound of farmed salmon; the ratio for cod is five to one, and for tuna is twenty to one, causing a clear ecological imbalance. Suggested improvements include replacing much of the fish protein source with chicken oil and soybean meal. Substitutes currently being investigated include protein made from grain and livestock byproducts, and omega-3 oils from single-cell microorganisms and genetically modified land plants. The Japanese group developing bluefin tuna aquaculture plans to gradually substitute plant protein for fish feed, but it is questionable whether tuna will take to this diet. Culturing herbivorous fish like carp appears to be much less environmentally damaging. Co-culture with seaweeds that could utilize the nutrients in fish waste seems to be an ecologically sound practice.

Increased efforts are going into offshore fish farms. Over the past decade the technology has steadily improved. Offshore fish farmers contend that by raising fish far from the coast they will avoid the environmental problems of near-shore aquaculture. Cages located miles offshore in deep water have less environmental impact because the ocean is vast and the water, wind, waves, and currents are constantly moving. Only a few studies have been done on offshore fish farming, but they've been positive. However, if deep-sea fish farming is to have any impact, it will have to get much bigger. Moving offshore also won't solve the problems of escapes or overuse of antibiotics, which can lower disease resistance, though it is possible that the vast open ocean will mitigate those effects.

Yonathan Zohar leads a group at the University of Maryland developing "green" aquaculture for branzini, European seabass. They make artificial seawater and recycle nearly all of it, filtering out waste and capturing methane to offset energy used in raising the fish. By maintaining high water quality the tanks are able to hold denser concentrations of fish than the typical open-water fish farm but without typical disease or parasite problems. They are also developing and testing new food made from plant material and algae, instead of fish meal.

The World Wildlife Fund (WWF) has organized "dialogues" among farmers, retailers, scientists, environmental groups, and other stakeholders to develop standards for certifying aquaculture products to minimize their environmental impacts. Criteria are being developed to conserve habitat, local biodiversity, and ecosystem function (associated with nutrient and chemical inputs), to protect the health and genetic integrity of wild populations (parasites, diseases, escapes), and to use resources and manage disease and parasites in an environmentally efficient and responsible manner (feed, chemical inputs like antibiotics). The U.S. Department of Agriculture has drawn up standards for aquaculture to be considered "organic." A technique for raising Pacific coho salmon in land-based freshwater tanks (avoiding escapes and sea lice issues) has been approved by the Monterey Bay Aquarium Seafood Watch Program. There are many more improvements in techniques and practices used in aquaculture that could result in it becoming a sustainable and beneficial industry that the world desperately needs in order to save the oceans.

TEN

Fish and Human Health

Question 1: Can eating fish improve your health?

Answer: Eating fishes can be healthy, which is one reason that so many fishes are being caught. Americans eat twice as much seafood as they did twenty years ago. Besides containing protein and nutrients such as vitamin D and selenium, fish contain omega-3 fatty acids that may reduce the risk of developing heart disease (the leading cause of death of men and women) and other medical problems. Fish oil contains both docosahexaenoic acid (DHA) and eicosapentaenoic acid (EPA). Fatty fish like mackerel, lake trout, herring, sardines, albacore tuna, and salmon are high in these fatty acids, which have been shown to reduce cardiovascular disease. Intake of recommended amounts of DHA and EPA in fish or fish oil supplements lowers triglycerides; reduces the risk of heart attack, abnormal heart rhythms, and strokes in people with cardiovascular disease; and slows the buildup of plaque ("hardening of the arteries"). Fish consumption is also linked to a lower risk of stroke, depression, and mental decline with age. The greater fish intake in the Japanese diet relative to that of the United States is associated with lower rates of heart attacks, heart disease, and plaque despite only moderately lower blood cholesterol levels. For pregnant women, mothers who are breastfeeding, and women of childbearing age, DHA is beneficial for brain development. Omega-3 fatty acids are also available in vegetable oils, walnuts, and yeast. Soybeans, canola, walnut, flaxseed, and their oils contain alpha-linolenic acid (LNA), which can become omega-3 fatty acid in the body. If more people got them from those sources, global fisheries might not be in such a critical state.

Question 2: Which species of fish are likely to be contaminated with pollutants?

Answer: Eating fish could have negative health effects because mercury and other pollutants accumulate in the food web. Fishes in polluted environments accumulate contaminants, but some have more than others. Larger, fattier, older fishes and those higher on the food chain generally have higher concentrations. Ironically, the fatty fishes that have more omega-3 fatty acids are also likely to contain more contaminants. States produce "fish advisories" that recommend either reduced (e.g., only one meal per week or per month) or no consumption of certain species from certain waters. In the spring of 2009 it was reported that striped bass and bluefish from the entire mid-Atlantic coast area had elevated PCBs.

Some fishes live a long time and over their lifespan can accumulate high levels of mercury and other contaminants, even when they don't live in a polluted habitat. The mercury may come from far away through the atmosphere; low amounts enter the ocean in rain. Algae take up small amounts, and when they die, sink, and are decomposed by bacteria, mercury is converted to the more toxic form, methylmercury, which gets more concentrated at each step of the food chain (*biomagnifies*). Many steps up the food chain later, predators like tuna and swordfish concentrate methylmercury from the fish they consume. They may have more than is considered safe by the Environmental Protection Agency (EPA) or the Food and Drug Administration (FDA) for women of child-bearing age (fetuses are at the highest risk from mercury, a neurotoxin), which makes them less appealing to everyone.

In January 2009, the FDA published a draft assessment of the benefits and risks of eating fish and concluded that mercury risks are small and that telling women to eat more fish has greater health benefits than telling them to eat low-mercury fish. The

FDA structured the choice as either "eat more fish" or "eat low-mercury fish," and in their scenario, fish benefits and mercury damage counteracted each other. They omitted a scenario that combines the two: "eat more fish, but only low-mercury fish." This would have greater benefits than doing one but not the other. The fishing industry praised the FDA analysis and urged that the current advisory be withdrawn or revised. The EPA, however, said FDA's analysis was flawed since they ignored advice from the National Academy of Sciences to use data from a certain study. EPA criticized FDA for many errors and faulty assumptions, including boosting benefits and reducing risks from seafood consumption beyond what was justified scientifically.

Question 3: Is there any value to taking cod liver oil?

Answer: Once used mainly as a bad-smelling vitamin supplement crammed down the throats of unwilling children by over-zealous parents, cod liver oil is now more likely to be recommended to older people suffering from osteoarthritis. There is evidence that it eases the pain of arthritis and improves joint stiffness. The main difference between fish oil and cod liver oil is that cod liver oil is very high in vitamin D, which is necessary in the winter when we don't get very much sun exposure on our skin in order to make vitamin D, which is important along with calcium in strengthening bones and combating osteoporosis. But cod liver oil can pose risks. While vitamin D may strengthen bones and lower the risk of developing cancer, one teaspoon of cod liver oil has 4,500 IU of vitamin A, and the standard dose is one to three teaspoons a day. Because cod liver oil has so much Vitamin A, it is possible to exceed the recommended dietary allowance (RDA). Excess vitamin A can interfere with bone growth.

Question 4: How can you determine that a fish you buy in the market is fresh?

Answer: There are two parts to this answer. There are things you can look for and there are labeling laws to help you choose.

If a fish has a strong "fishy" smell, pass it up—it probably is not fresh. Pull open the gill cover—if the tissue underneath is bright red, it is probably fresh. The eye should look healthy and clear, and the scales should be even and shiny.

The U.S. Department of Agriculture has regulations that require seafood to have the country of origin labeled. This should enable you to avoid fish from a country known to have problems with pollution or quality control. Stores are not required to label fish if it has been processed in any way, and that includes marinated fish. The U.S. Food and Drug Administration requires that farmed salmon, often fed additives that turn the flesh pink, be labeled "color added." They also require that a fish that has been thawed before being put out for sale be labeled "previously frozen." Surveys indicate that implementation of these regulations is spotty. Try to find a fish store that follows the rules.

Question 5: Which kinds of fishes might be toxic if you eat them?

Answer: Many species of pufferfish (see color plate H), family Tetraodontidae, contain a toxin called tetrodotoxin, which is concentrated in the internal organs like gonads, liver, and intestines, but may be found in smaller concentrations in muscle tissue. These fishes, which inflate themselves when threatened, are very tasty. The first recorded cases of tetrodotoxin poisoning were from the logs of Captain James Cook in September 1774. He recorded that his crew ate some pufferfish and then fed the remains to the pigs on board. The crew experienced numbness and shortness of breath, while the pigs (having eaten organs) were all dead the next morning.

Fugu puffers, *Takifugu*, from Japan are toxic, but are nevertheless popular gourmet foods. Special chefs at restaurants prepare and cook the fish so that, ideally, the diner will experience a tingling sensation, blurry vision, and other pleasant sensations of intoxication, but nothing more serious. This goal is not always realized, however, as puffer poisoning cases, some fatal, still occur. If the chef miscalculates a person's resistance to the toxin, paralysis and death can result. Up to two hundred cases annually are reported, with mortality approaching 50 percent. Why do the Japanese do this? Probably for the same reason as people go sky-diving. An aquaculture company in Japan says it raised fifty thousand fugu that are nonpoisonous. It will be interesting to see whether they become popular in Japan. The puffer in the northern Atlantic, *Spheroides maculatus,* fortunately for people in that area, is of low toxicity, and its muscular abdomen is marketed as "sea squab," a delicacy.

Tetrodotoxin, one of the most potent toxins known, causes a failure of nerve conduction, particularly at the nerve-muscle junction, causing paralysis. Initial symptoms are numbness and tingling of the lips, tongue, and inner surfaces of the mouth (probably the reason for eating fugu), but this is often followed by weakness and paralysis of the limb and chest muscles and a drop in blood pressure. Death can occur within thirty minutes because the victim can no longer breathe. Until recently, tetrodotoxin was assumed to be produced by the puffers themselves, but it is probably produced by marine bacteria. Tetrodotoxin-producing bacteria have been found in puffers. It appears that the fish, which have evolved resistance to the toxin, become toxic by storing it in their tissues. It is found also in the skin of some newts and frogs and is a useful defense for them as well as for the puffers.

Fish eaters may also be poisoned by eating marine fish containing a toxin called *ciguatoxin,* produced originally by the single-celled dinoflagellate *Gambierdiscus toxicus* in the Caribbean and tropical Pacific Ocean. This toxin *biomagnifies* through the food chain, so that species higher in the food chain, like barracudas, snappers, moray eels, groupers, and amberjacks, accumulate the most. Symptoms of *ciguatera* include nausea, vomiting, and diarrhea, followed by headaches, muscle aches, numbness, poor coordination, and hallucinations. Most people recover within a few weeks, though symptoms may persist for years. The presence of the toxin is very spotty, so not all fish of the same species caught at the same time and same place will be toxic. The only way to be sure to avoid exposure to the toxin is to avoid consumption of large reef fishes. Ciguatoxin is heat-resistant, so fish cannot be detoxified by cooking.

Question 6: Is sushi safe to eat?

Answer: If fish is eaten raw, consumers may acquire parasites. These are mostly worms—nematodes (round worms), cestodes (tapeworms), and trematodes (flukes). Some trematodes that can be transmitted from fish to people are intestinal parasites, while others are liver parasites that can cause more serious damage. There are over fifty different worms that can infect humans from uncooked fish. However, for sushi and sashimi lovers there is good news: parasitic infections from marine fishes are rare. Parasites are more likely in fishes that spend time in brackish or fresh water like Pacific salmon, most of which were found to have roundworm larvae in their flesh (the type that might infect humans). Pelagic fishes like tuna usually have few parasites due to their extensive migrations; they are not in any one area long enough to eat many prey that might have a lot of parasites. There are, however, reports of humans becoming infected with roundworms after eating raw yellowfin tuna and with tapeworms after eating raw salmon. Salmon for sushi and sashimi should be frozen first to kill any parasites. A parasitologist used to give a seminar entitled "If You Knew Sushi like I Know Sushi," which would discourage most people. I confess to still eating sushi, but not too often. However, there is still the issue of mercury in the tuna to consider. Actor Jeremy Piven left a Broadway show in the spring of 2009, claiming to have mercury poisoning from eating too much sushi (apparently two meals a day).

Question 7: What kinds of fishes should we (or should we not) eat?

Answer: There are different reasons why eating certain fish may be discouraged. One reason is because long-lived species can accumulate contaminants like mercury. Sharks, swordfish, some bluefish, some striped bass, and some tuna may have more mercury than recommended for women of child-bearing age and small children. For freshwater species, states have advisories against eating predatory fish—typically, large-mouth bass and pickerel—from specific areas where mercury is elevated.

Another reason to be cautious about consuming fish relates to concerns about overfishing. The more the market demand for a species, the more fishing will take place. Species of concern include bluefin tuna (see fig. 26, in chapter 9), Atlantic cod (see fig. 19, in chapter 5), sturgeon, swordfish, and Chilean sea bass (previously called Patagonian toothfish). It is interesting to note that several species have been renamed to make them more appealing—the overfished orange roughy used to be called "slimefish." Eating tuna and swordfish is discouraged not only for ecological reasons but also because they are on the "mercury list." Ecological concerns about eating certain species relate to more than just their scarcity. A fishery may damage the environment by bottom trawling or may have a lot of bycatch. Salmon aquaculture can deplete oxygen, contaminate the environment with antibiotics and anti-parasite chemicals, and may transmit parasites to wild salmon (see chapter 9: "Commercial Fishing"). About seventy species of fishes have been banned from menus at thousands of restaurants across the United Kingdom and Ireland. The Compass Group, the world's largest contract caterer, has decided that these fish should not be eaten because of environmental concerns.

A number of organizations have published guides to "smart" seafood, in which species that are fished or raised in sustainable ways are the "best" or "green" choices; those with some concerns about how they're fished or their health are in yellow, and those that should be avoided (overfished or fished or farmed in ways that harm the environment) are in red. These lists may vary a bit, but are pretty similar. Lists of good and bad choices, which are available from Food and Water Watch, the Environmental Defense Fund, and the Monterey Bay Aquarium, among others, also include shellfish. The major North American seafood certification groups have come together in a Web site called fishchoice.com, where buyers can select which certification scheme

Green, Yellow, and Red Lists

Various organizations have developed lists of seafood choices that are good for you (low in contaminants) and good for the ocean—with sustainable fisheries. The species on the green list are the best choices, those on the yellow list are the OK choices, and those on the red list are the ones to avoid because of health or ecological reasons. The list from the Blue Ocean Institute (http://www.blueocean.org/seafood/) is as follows:

Green: Atlantic herring, walleye pollock, Atlantic mackerel, king mackerel, Spanish mackerel, cero mackerel, chub mackerel, striped bass, wahoo, tilapia (farmed), Alaska salmon, weakfish, mahi-mahi (pole and line caught), rock sole, yellowfin sole, yellowfin tuna (pole and line or purse seine caught), albacore tuna (pole caught), skipjack tuna (pole and line caught or purse seine caught), sablefish (Alaska), black sea bass, catfish (farmed), hybrid striped bass, lingcod, summer flounder, Dover sole, Pacific cod, Pacific halibut, and Petrale sole.

Yellow: Canned tuna, mahi-mahi (longline caught), Pacific salmon (WA, OR, CA), rainbow trout (farmed), swordfish (Pacific), bluefish, monkfish, yellowfin tuna (longline caught), rockfish (Alaska), albacore tuna (longline caught), and sharks.

Red: Atlantic flounders and soles, hoki (New Zealand), Icelandic cod, steelhead, American eel, rockfish (U.S. West Coast), Atlantic bluefin tuna, Chilean sea bass, groupers, orange roughy, Atlantic cod, Atlantic halibut, oreos, sturgeon caviar (U.S.), Caspian Sea caviar, snappers, and Atlantic salmon (farmed).

they wish to follow and see the contact information for suppliers who can sell that fish.

The location and way in which the fish were caught can be important, and consumers should ask restaurants, fish markets, or grocery stores about the origin of the fish they are selling.

Unfortunately, sellers frequently don't know where their fish were caught or how. Consumers should balance risk and benefit when choosing seafood. There are many tasty and healthy choices. Choose seafood low on the food web, short-lived fish, those high in omega-3s, and select a variety of species from different locales. Try to choose fish caught or farmed sustainably, perhaps by looking for the Marine Stewardship Council label (see chapter 11, question 6: What can I do to help protect fishes?) or shopping in stores that label their seafood. Unfortunately, some of the most abundant (green list) fish like herring are difficult to find. Furthermore, when you find a fish you think is abundant, the terminology can be confusing. Seafood Watch names "black sea bass" a "good" alternative, but mentions that several other species are called by similar names, including endangered groupers. Consumers should also be aware that there have been many cases of mislabeling—the fish is a different species from what it is supposed to be. More than one-quarter of retail fish samples analyzed for DNA in Vancouver, Canada, turned out to be mislabeled. Tests at sushi restaurants produced similar results. It is also paradoxical that frozen fish may have less environmental impact than fresh fish. When fresh fish are driven a reasonable distance to market, the relative environmental impact is low. But they are generally flown long distances. When fish are flash-frozen at sea, the taste and quality are equivalent to fresh and they can be moved thousands of miles by container ship, rail, or truck with much less environmental impact than fresh fish that are air freighted. There is no easy solution other than not eating fish very often, which may help fisheries recover.

Question 8: How are fishes used in traditional Eastern medicine?

Answer: In traditional Chinese medicine, the seahorse (see color plate D) is in great demand. Indeed, the Chinese use more than 20 million a year. After being dried, seahorses are used to treat a variety of disorders, including asthma, arteriosclerosis, incontinence, impotence, thyroid disorders, skin ailments, broken bones, heart disease, as well as to facilitate childbirth and even as an aphrodisiac. Seahorses are also used as medicines by Indonesians, Filipinos, and others. The best seahorses in traditional Chinese medicine sell in Hong Kong for up to 550 U.S. dollars per pound. There are about forty countries involved in the seahorse trade. The practice of using endangered species is controversial since it is having a major impact on the populations of seahorses. TRAFFIC, the international agency that monitors the trade in endangered species, started a dialogue and arranged a workshop with the traditional medicine community and conservationists. They found that once discussion began, the Chinese medicine community was receptive, so there may be some hope for survival of the seahorses.

Shark fin soup is traditionally regarded as beneficial for health in East Asia, and its status as an "elite" dish has led to huge demand with increased affluence in China. Shark fin soup is regarded as a tonic for strengthening the waist, supplementing vital energy, nourishing blood, invigorating kidney and lung, and improving digestion. Nutritionists find it rich in protein, and the gelatin can help the growth of cartilage. Many species of sharks are used for their fins. The United Nations Food and Agriculture Organization says over 100 million sharks, skates, and rays are killed every year. They are caught, their fins removed, and the fish discarded back into the ocean to die. This practice is having a devastating effect on shark populations because they reproduce very slowly. Shark populations are declining rapidly, some species by 90 percent.

ELEVEN

Research and Conservation

Question 1: Why do people study fishes?

Answer: It is likely that the earliest ichthyologists were people who had learned how, where, and when to obtain the most useful fish. But beyond that practical knowledge, fishes are so varied and do many interesting things that scientists want to learn about. Fishes are excellent subjects for studying physiology, anatomy, behavior, development, ecology, and evolution. I asked several fish researchers why they decided to study fishes and got a variety of answers. Rebecca Jordan originally wanted to study dolphins and took aquatic-related courses in college including Ichthyology. But it wasn't until she was introduced to cichlids and their complex behavior that her interest was piqued: "The capacity of fish to recognize each other, to make decisions based on associative memory, or to learn in general fascinates me!" Howard Reisman says: "Like many vertebrates, fishes have evolved clever ways to solve the problems of remaining alive and producing young. However, because there are more fishes than all of the other vertebrates combined, living under a variety of daunting conditions, fishes are pressed to develop even more intriguing solutions." Francis Juanes intended to study invertebrates but moved into fishes because they are abundant and are amazingly variable in their life histories: "They display huge ranges of body sizes even within species and can have large effects on their environments, their predators and prey." Peter Moller was fascinated by fishes that produce electricity for communication and orientation, and how their various senses work together. Also, studying African mormyrid fishes gives him occasions to explore them in their natural habitats in West and Central Africa. Walter Courtenay was fascinated the first time he caught a walleye pike at about age seven or eight. He then spent many summers fishing "but spending much time along a creek just watching fishes in those (then) clear waters. Opposite of what I had been doing to fishes by catching them, they caught me!"

Certain species are particularly popular for certain kinds of studies. The dogfish shark and yellow perch are often used in the study of anatomy in undergraduate courses. The fathead minnow, *Pimephales promelas,* and mummichog, *Fundulus heteroclitus* (see fig. 16, in chapter 4), are commonly used to study effects of water pollution, while the medaka, *Oryzias latipes,* and zebrafish, *Danio rerio* (fig. 28), are often used in studies of development. Fishes may be used as "models" to learn about vertebrates in general, including humans, or the scientists may be interested in the fishes themselves and want to learn about their anatomy,

Figure 28. Zebrafish, *Danio rerio. (Photo by Steve Baskauf.)*

biochemistry, physiology, development, behavior, or ecology, either for its own sake or with a goal of improving fisheries management or aquaculture.

Question 2: How are zebrafish used as a model for genetic research?

Answer: Zebrafish, *Danio rerio* (fig. 28), native to freshwater streams in India, have been popular aquarium fish for many years. In the early 1970s, Dr. George Streisinger at the University of Oregon found that they are a good model for studying vertebrate development and genetics. Since then, zebrafish embryos have become popular as a model for understanding embryonic development of not only fish, but vertebrates in general, including people. Zebrafish are easy to keep and reproduce readily in the laboratory. Their eggs are transparent, allowing scientists to watch them under a microscope develop in a few days into a newly formed fish (see fig. 20, in chapter 5). Scientists can manipulate the embryos and move a cell to another location to see if it will still develop the same way as it does normally or if it will do something different. This provides insight into the mechanisms of development and may be relevant to birth defects in humans. As vertebrates, zebrafish are much closer to us genetically than fruit flies, roundworms, and bacteria, other species commonly used in genetic research. Large-scale chemical mutagenesis experiments isolated about two thousand mutations affecting all aspects of embryonic development. Once the mutations were identified, their effects could be seen, and their genetic and molecular nature determined. In the 1990s, the entire zebrafish genome was worked out. Its usefulness as a model is due to its genetic and embryological characteristics. Strains have been developed that enable scientists to understand the relationship between certain genes and certain abnormalities so that they can identify what the gene does. Genetic defects in zebrafish frequently resemble human disorders, so they can be used to study basic biological questions and as a model for human inherited diseases. Because of its stripes, genes involved in pigmentation were studied and found to be the same as in humans. Recently, a strain with transparent bodies was developed, which allows viewing of blood cells and metastasizing (spreading) cancer cells. Since many genes are the same in fish and humans, this strain may provide insight into human cancers.

The most common cause of hearing loss in humans is damage to hair cells in the inner ear. The hairs (which cannot regenerate) bend in response to sound waves, triggering the cell to send an impulse to the brain. Zebrafish lateral lines have hair cells that can regenerate; they are being studied with the goal of understanding mechanisms to protect hair cells in our inner ears and to stimulate our hair cells to regenerate.

This celebrated species now has a whole scientific journal devoted exclusively to it. It is being used in schools from elementary through college to teach developmental biology, genetics, neurobiology, and behavior. The embryos are easy to maintain, and embryonic development can be observed with common classroom equipment.

Question 3: Are any fishes endangered or recently extinct?

Answer: It is hard to imagine that our oceans, lakes, and rivers—underwater worlds so abundant with life—could one day be without some of the familiar fishes that call them home. But it could happen within our lifetime. When fishes are caught faster than they can reproduce or are damaged by pollution or habitat loss, not only is the health of aquatic systems jeopardized, but also the livelihoods of fishermen and the communities that depend on them.

The American Fisheries Society (H. L. Jelks and others) in 2008 provided a detailed evaluation of the conservation status of seven hundred listed species of North American freshwater fishes and diadromous fishes (which migrate between rivers and oceans). They reported that nearly 40 percent of the North American freshwater fishes are classified as in jeopardy; 230 species are considered vulnerable, 190 are threatened, 280 are endangered, and 61 are presumed extinct. Of those that had been imperiled twenty years earlier, about 90 percent were the same or worse. The causes of this decline include habitat loss, reduced range, introduction of non-native species, and climate change. The problem is most severe in the southeastern United States, the mid-Pacific coast, the lower Rio Grande, and Mexican basins that do not drain into the sea. This reflects worldwide trends in the loss of freshwater fishes, many of which have narrow ranges and exist as small, isolated populations that are particularly vulnerable to habitat loss. Although historical records are scarce, freshwater fishes appear to have been superabundant in North America during the early seventeenth century. Settlers quickly learned to exploit the fishes, including migratory species and lake residents like the lake sturgeon, *Acipenser fulvescens,* which was the most important commercial species in the Great Lakes until the 1920s, when the population collapsed and commercial fishing was banned. The future appears bleak for North American and Australian freshwater fishes, especially those that have been commercially harvested. Since most commercial species are large predators, their removal causes changes in food webs and nutrient dynamics, just like in the oceans.

Marine fishes have experienced severe population declines, but few species in U.S. waters are at the point of being endangered or extinct. However, oceanic species beyond the control of individual countries are endangered. In developing countries the situation can be severe. Over eighty species have either become extinct or are missing from Malaysian waters, according to S. M. Mohd Idris, president of an environmental organization. Describing the situation as alarming, he urged effective fishery management, enforcement of existing laws, and preservation of important habitats like mangroves, sea grass, and coral reefs.

Migratory fishes are also in trouble from dams, habitat loss, overfishing, and pollution. Pacific salmon in the lower forty-eight states are in such danger that they have been listed as threatened or endangered. They are doing reasonably well only in Alaska, but are threatened by mining and logging, which disrupt the streams where they breed. The shortnose sturgeon,

Acipenser brevirostris (fig. 29), was in low numbers in the Hudson River in the 1960s, but the Endangered Species Act prohibited fishing and ordered that pollution be cleaned up from spawning grounds near Albany. Today the Hudson River population is healthy, but other populations of these ancient fishes are not so lucky. The Atlantic sturgeon, *A. oxyrhinchus* (fig. 29), remains endangered despite closure of the fishery (for caviar) some years ago. It was a major fishery one hundred years ago, with annual harvests approaching 7 million pounds. In the late twentieth century, 100,000 to 250,000 pounds were caught annually until a moratorium was implemented in 1997. Some sturgeons in the Pacific Northwest are showing slow recovery, but it is very slow since they may not mature until they are twenty years old, may not breed every year, and are vulnerable to fishing and pollution. Sturgeons from the Black and Caspian seas are also endangered for the same reasons, with some populations nearing extinction. In Chesapeake Bay and the Hudson River alewife, *Alosa pseudoharengus;* American shad, *A. sapidissima;* and blueback herring, *A. aestivalis* populations have dwindled to about

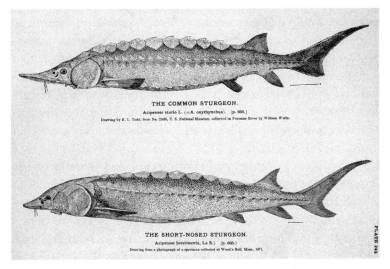

THE COMMON STURGEON.
Acipenser sturio L. (=A. oxyrhynchus). (p. 660.)
Drawing by H. L. Todd, from No. 22495, U. S. National Museum, collected in Potomac River by William Wolfe.

THE SHORT-NOSED STURGEON.
Acipenser brevirostris, Le S.). (p. 660.)
Drawing from a photograph of a specimen collected at Wood's Holl, Mass., 1871.

Figure 29.   Common (Atlantic) sturgeon, *Acipenser oxyrhinchus,* and shortnose sturgeon, *A. brevirostris. (Courtesy of NOAA.)*

10 percent of what they were formerly, due to dams and large numbers being caught as bycatch in ocean fisheries. Eels are also in trouble. Norwegian regulators in a landmark decision have banned all fishing of the critically endangered European eel starting in 2010 and cut 2009 quotas by 80 percent. Dam removal can have very positive effects. Ten years after the Edwards dam was demolished, the Kennebec River in Maine has seen fish return to spawn and thrive. Reduction of pollution has allowed the return of Atlantic salmon and shad to the Seine River in France, where they had been locally extinct for many decades.

Question 4: Why have so many fish populations crashed despite local, national, and international regulations?

Answer: While each country regulates fishing within a 200-mile limit, oceanic swordfish, sharks, and tunas are not protected by the laws of any country and must be protected by international agreements that are seldom driven by science, effective, or enforced. Scientists at the International Commission for the Conservation of Atlantic Tunas (ICCAT) in 2007 recommended a bluefin tuna catch of 10,000 to 15,000 tons per year, but European Union (EU) ministers set a quota of 29,000 tons, ignoring scientists' advice. In fact, the amount of bluefin landed was 61,000 tons—four times what scientists had recommended—due to illegal and unreported fishing. Callum Roberts of York University says that politicians should decide only how the catch should be allocated among different nations but not how big the catch should be in the first place. That should be set by scientists.

Many sharks are threatened with extinction. A study of the status of 64 species of open ocean sharks and rays revealed that one-third are threatened with extinction primarily due to overfishing, according to the International Union for the Conservation of Nature. The great hammerhead, scalloped hammerhead, and giant devil ray are globally endangered. Smooth hammerheads, great white, basking, and whitetip sharks are globally vulnerable to extinction, along with two species of makos and three species of threshers. Finning—cutting fins off a living shark and dumping the carcass back into the water—is a particularly difficult problem, and some nations are banning it. The Shark Conservation Act would require all sharks caught in U.S. waters to be landed with their fins still attached. EU countries are the main exporter of shark fins to China for shark-fin soup. In April 2009, EU ministers drew up an action plan to restrict finning, a wasteful practice that discards over 90 percent of the shark.

Fishes are in trouble due to many factors besides fishing: habitat alteration, pollution, climate change, and the introduction of non-native species. The aquatic environments most at risk from non-fisheries issues are lakes, big rivers, and small isolated waters, such as those inhabited by desert pupfish. Lakes are particularly susceptible to invasive species, while rivers are dammed, channelized, and polluted. The Colorado River is home to many species that live nowhere else but are disappearing because of drastic modifications to the river, which has been divided by dams into a series of reservoirs separated by regulated streams. The Glen Canyon Dam creates unnatural ebbs and flows, making a poor environment for chubs, squawfish, and suckers that are adapted to free, fast-moving waters, so they are nearing extinction. However, in June 2009 a federal court told the Bureau of Reclamation, which operates the dam, to change its dam operations to make them less harmful to species like the chub. The Great Lakes, too, have been affected by overfishing and urban pollution. Five native species that once supported active fisheries have gone extinct. Non-native sea lampreys, smelt, and alewives also contributed to the decline: the lamprey fed on them, while the others outcompeted them for food.

Climate change has and will continue to have negative effects on fish populations. Walleye in Lake Erie survive better when the water is cool (around 34 degrees Fahrenheit), as it is in March at the beginning of the spawning season. Ten degrees higher, and the populations decline. With rising temperatures, the number of these important game fish will dwindle. Similarly, in winter, when the ocean's temperature dips below 68 degrees Fahrenheit, millions of sardines migrate over 1,000 miles along the east coast of South Africa (see fig. 14, in chapter 4). This spectacular sardine run with schools miles long attracts predators that depend on this source of food. The best sardine runs happen when the water is cold; fewer fish make the journey as ocean temperatures rise.

Introduction of non-native species can have devastating effects. Lake Victoria (or Victoria Nyanza) in East Africa is a large, deep rift lake that used to support a vast collection of endemic fishes, mostly cichlids. Hundreds of different species of cichlids specialized into many diverse niches. When predatory Nile perch, *Lates niloticus,* were introduced as a fishery in the 1950s, they ate and wiped out many of the cichlids. It is estimated that over two hundred cichlid species have gone extinct.

Question 5: What are some new ways to protect fishes and their ecosystems?

Answer: Fisheries need to be reformed in a number of ways, according to Callum Roberts of York University in England: reduce present fishing capacity, eliminate risk-prone decision-making, eliminate catch quotas and instead control the amount of fishing, require people to keep what they catch, require fishers to use gear modified to reduce bycatch, ban or restrict the most damaging methods, and implement networks of marine reserves that are off-limits to fishing.

Traditional fishery management sets an overall limit on how much fish may be harvested, which results in fierce competition among fishermen to catch the greatest amount in a short period. They go out over a few chaotic days and take risks, such as going out in stormy weather, in order to catch the most fish within the limits. The system provides no incentives for conservation or stewardship of the resource. An alternative management technique is to allocate individual "catch shares" or rights to a percent of the harvest to individuals or groups. Allocating shares of a fishery to individuals, cooperatives, or communities gives them incentives to help the stock grow. In this system, scientists set catch levels, and other authorities allocate shares, species by species, region by region. As a local stock grows, catch shares or individual transferable quotas become more valuable, just as shares of a company's stock become more valuable as the business prospers. Those who have shares have an incentive to protect the stock. Each year before the season begins, fishers know how much they are allowed to take of the total allowable catch, a number set each year by scientists after a fishery stock assessment. Like shares of stock in a corporation, catch shares (individual transferable quotas or ITQs) can be bought and sold. Each share becomes more valuable when the fish population increases. Fishermen are motivated to maximize the value of their shares by having the fish population increase, a clear economic rationale for conservation. As the fishery becomes more efficient and less competitive, fewer boats and gear are needed and fishermen can plan more effectively, delivering fish to meet market demands and staying ashore in bad weather.

Catch-share fisheries have had far fewer collapses than traditionally managed fisheries. It has been estimated that if catch shares had been in place globally in 1970, less than 10 percent of the world's fisheries would have collapsed by 2003, rather than 27 percent. The Pacific halibut fishery is an example of success. In 1995, the fishing season had dwindled from four months down to just two or three days when fishermen would go out in any weather and stay till it closed. Boats would make dangerous sprints and risk colliding with each other, with holds crammed full of frozen fish. Some boats sank from being overloaded. Then they converted to ITQs and the season lasts nearly eight months. Because boats haul in fresh, undamaged fish in manageable quantities, the price the fishermen get has increased significantly. In April 2009 the EU Commission recommended that EU states adopt an ITQ system, since 90 percent of stocks were exploited beyond their ability to recover.

Of course, catch-shares can only be effective if fisheries managers set the allowed catch at a level that is scientifically based on the estimated size of the fishery. But fish recruitment fluctuates based on environmental conditions. Fishery managers are under pressure not to lower the allowable catch, even if the

long-term survival of a fishery depends on reducing (or elimi-
nating) fishing. The levels of uncertainty involved in estimating
fish stocks make defendable decisions very difficult, given the
economic impacts. This new approach, that transforms commonly held resources into government-granted and economi-
cally tradeable fish-catching quotas, is controversial, and many
fishing organizations are still opposed to it. They fear that small
fishing operations will be forced to close and that without buy-
outs there will be severe economic consequences for small boat
operations.

Another relatively new approach to protecting fishes and their
ecosystems is setting aside Marine Protected Areas (MPAs) in
which no fishing is allowed. This is nothing new on land—we
have had National Parks with no hunting for over a century—
but it wasn't until late in the twentieth century that the idea of
MPAs was taken seriously. MPAs can be established for many
reasons: to protect a certain species, to benefit fisheries management, or to protect full ecosystems, rare habitat, or nursery
grounds. MPAs can be very large (Great Barrier Reef) or very
small. They may be temporary, to protect vulnerable stages in
the life cycle like spawning grounds and nursery areas. A network of reserves may be developed to protect different habitats,
species, and migratory routes. Reserves allow habitats to be re-
stored and benthic species such as corals, sponges, and mollusks
to recover from trawling damage.

Well-managed MPAs generally show an increase in numbers
and diversity of marine life and improvement of overall ecosystem health. The quantity of commercially important fish inside
a reserve can double within a few years. Protection from fishing
allows fish to live longer, grow bigger, and become more abundant. Bigger females produce more and better eggs. Benefits
are seen within the reserve, but since the fishes are unaware of
the boundaries, some move out, increasing the populations
outside the reserve. If areas with older fish are protected, they
provide "spillover" of both offspring and mature fish, increasing the potential catch for the fishery outside the boundary.
A reserve in Apo Island in the Philippines produced a tenfold

increase in the nearby fishery in twenty years. After haddock crashed in New England in the 1990s, over 20,000 square kilometers around nurseries and areas of high abundance were closed. After ten years, haddock showed a dramatic recovery. The benefits of MPAs include maintaining biodiversity and providing refuges, protecting important habitats from damage by destructive fishing practices, allowing damaged areas to recover, providing areas where fish can spawn and grow to adult size, increasing fish catches (both size and quantity) in surrounding fishing grounds, and building resilience to protect against damaging external impacts like climate change. By maintaining viable fisheries, reserves can help to maintain local cultures, economies, and livelihoods linked to the health of the marine environment. Particularly effective are networks of reserves in which larvae produced in one reserve can disperse and find their way into neighboring reserves. A recent study using DNA analysis by Sergio Planes of Perpignan University in France showed that while about 40 percent of anemonefish juveniles settled at the same reef where they were spawned, others moved up to 35 kilometers away, in some cases into other reserves—a new way of "finding Nemo."

MPAs not only help protect biodiversity and demonstrate that the seas can recover from overfishing, but also benefit fisheries and people. Several countries, including South Africa and Australia, have made substantial progress in setting aside reserves. The United States recently set aside valuable areas, including the Northwestern Hawaiian Islands Marine National Monument and grouper spawning aggregations in the Atlantic and Gulf coasts. However, less than 1 percent of the oceans are protected, and the vast majority of existing marine parks and reserves have little or no effective management.

In fresh water most of the efforts to protect fishes have been habitat restoration projects such as removing dams, cleaning up pollution, and restoring wetlands. There also have been attempts to re-introduce species that have disappeared. An approach similar to MPAs that is less well developed is the idea of establishing protected freshwater areas.

Question 6: What can I do to help protect fishes?

Answer: There are many reasons to care about fishes—ethical, aesthetic, and practical. Ethically, it isn't right to eliminate other species from this planet (except perhaps some diseases). Aesthetically, each species is unique, with characteristics that have enabled it to survive millions of years. Practically, fish may be a resource as food or something else, and they play a role in the ecosystem they inhabit. Removing species from an ecosystem weakens its overall functioning.

Citizens can participate in political decision-making and write letters to their representatives or testify at hearings. They can join groups interested in fish conservation and learn more about the issues. Some major organizations that are actively working on general conservation (including fish) are the World Wildlife Fund, the Nature Conservancy, National Resources Defense Council, Sierra Club, and Environmental Defense Fund. Some groups focus particularly on aquatic or marine habitats, including the Ocean Conservancy, American Littoral Society, and Oceana. Some focus exclusively on fishes, for example, the North American Native Fishes Association, the Marine Fish Conservation Network, and Fish Unlimited. Some focus on local areas, for example, the network of Riverkeeper and Baykeeper organizations, each of which focuses on a particular water body. Some organizations focus on a particular fish, such as Trout Unlimited, Project Seahorse, the Desert Fishes Council, and the World Sturgeon Conservation Society. Some organizations are more militant than others. At the European Seafood Exposition in 2007, Greenpeace activists managed to close down exhibitors trading in critically endangered bluefin tuna by draping their stands in fishing nets, chaining themselves to fixtures, and putting up banners that read, "Time and tuna are running out." Their main target was the Mitsubishi Corporation, the Japanese car manufacturer that is also the world's largest tuna trader, controlling 60 percent of the market and accounting for 40 percent of all bluefin tuna imported into Japan from the Mediterranean. Greenpeace also sank over one hundred huge boulders into protected cod fishing grounds off Sweden to impede bottom-trawling.

If you have a home aquarium, you can ascertain that the fish you buy are cultured rather than caught from the wild with destructive practices. (This will likely mean having a freshwater rather than reef tank.) Overfishing is a concern. The Banggai cardinalfish is found only in a few areas off the coast of Indonesia and has been labeled an endangered species by the International Union for Conservation of Nature, largely because of overharvesting for the aquarium trade. If you live near one of the Marine Aquarium Council–certified retailers in the United States, you can buy marine fish that are verified to have been collected sustainably or cultured and properly handled throughout the supply chain.

You can also order something else when endangered or threatened species (e.g., orange roughy, Chilean sea bass, swordfish) appear in a fish market or on a menu, and speak to the management. The Marine Stewardship Council (MSC) is an international organization that created standards for sustainable fisheries in the late 1990s. See if fish in your market are certified by MSC. The label can make a big difference: 42 percent of the world's wild salmon catch is MSC-certified. About 5 million tons of seafood is certified by the MSC, only 5 percent of wild-caught seafood. However, supermarkets like Whole Foods have listened to the MSC and have developed their own policies. They can buy or not buy from a particular supplier and can wield a lot of power. Ultimately, an informed and concerned fish-eating and fish-watching public can make a huge difference in the survival and well-being of these wonderful creatures.

Appendix: Public Aquaria in the United States

Following is a list, alphabetical by state, of many of the public aquaria in the United States. They all have Web sites with additional information.

Alabama

The Estuarium at Dauphin Island. It has exhibits of the Mobile-Tensaw River Delta wetlands as well as the Mobile Bay estuary.

California

Birch Aquarium at Scripps, La Jolla. Focused on science education, it has two seawater-equipped classrooms next to public display areas.

Long Beach Aquarium of the Pacific. One of the largest in the United States, its exhibits take visitors throughout the Pacific Ocean's three regions: Southern California/Baja, the Tropical Pacific, and the Northern Pacific.

Monterey Bay Aquarium. Exhibits re-create Monterey Bay's habitats, from shallow tide pools to the deep sea. It is one of the best aquaria in the nation. The Web site offers webcams, photo galleries, and a wealth of information. There is a top-notch research facility that has developed techniques of culturing marine animals such as jellyfish, which now are exhibited in many other aquaria.

Sea World, San Diego. You can travel through a submerged viewing tube, a pipeline to the center of the shark's world, one of the largest displays of sharks.

Steinhart Aquarium, San Francisco. A new facility opened a few years ago and offers more displays than the former site. Exhibits are both salt water and fresh water.

Colorado

Downtown Aquarium, Denver. This re-creates habitats of the Colorado River as it flows to the Sea of Cortez and of Indonesia's Kampar River along its journey to the South China Sea.

Connecticut

Maritime Aquarium, Norwalk. Devoted to the Long Island Sound, it features over one thousand animals native to the Sound and its watershed and has an IMAX theater.

Mystic Aquarium and Institute for Exploration, Mystic. In addition to the many exhibits, it has educational programs for different ages.

District of Columbia

Smithsonian National Zoological Park. It has a variety of exhibits, including an Amazonian fish display.

Florida

Florida Aquarium, Tampa Bay. It focuses on Florida as it follows a drop of water from its underground source to the open sea. Exhibits feature thousands of native plants and animals as well as presentations from around the world.

Sea World, Orlando. It has exhibits where you can feed the dolphins. The site sports an interactive map so you can check out the exhibits before you visit.

Hawaii

Waikiki Aquarium. The third oldest public aquarium in the United States, it is a small aquarium with over three hundred species of aquatic animals and plants, including endangered and threatened species.

Illinois

Shedd Aquarium, Chicago. Although many of the exhibits are salt water, the Amazon Rising: Seasons of the River exhibit is an excellent freshwater display.

Kentucky

Newport Aquarium. Sixteen galleries each present a different theme. It offers group rates and student programs.

Louisiana

Audubon Institute Aquarium of the Americas, New Orleans. On the Mississippi Riverbank, it hosts daily "expeditions" into various aquatic habitats that include thousands of fish, reptiles, and birds, including many native to Louisiana.

Maryland

National Aquarium, Baltimore. It is sited on a tributary of the Chesapeake Bay, the nation's largest estuary, and is home to over ten thousand marine and freshwater animals in many exhibits.

Massachusetts

The Berkshire Museum, Pittsfield. Exhibits feature freshwater fish of the Amazon, among others. A number of educational programs are offered.

New England Aquarium, Boston. It has a large variety of exhibits and an IMAX theater.

Minnesota

Great Lakes Aquarium, Duluth. The only all freshwater aquarium in the United States, it houses seventy species of fishes.

Underwater World, Bloomington. Below the Mall of America, this aquarium features a variety of exhibits.

New Jersey

Adventure Aquarium, Camden. It has a wide range of freshwater and saltwater exhibits. Freshwater displays include South American electric fish, Central American blind cavefish, and Asian bumblebee fish.

New Mexico

Albuquerque Aquarium. Exhibits encompass habitats of the Gulf of Mexico. It is open summers only.

New York

Atlantis Marine World, Riverhead. In addition to some freshwater and mostly marine exhibits, it has the largest living closed-system coral reef display in the Western Hemisphere (20,000 gallons) and a 120,000-gallon tank with sharks, groupers, and turtles.

Cold Spring Harbor Fish Hatchery and Aquarium, Cold Spring Harbor. It features New York State freshwater reptiles, fishes, and amphibians and presents many educational programs.

New York Aquarium, Coney Island, in Brooklyn. It is home to thousands of fishes and other marine animals, including beluga whales, walruses, and dolphins.

Oregon

Oregon Coast Aquarium, Newport. One of the largest aquaria in the country, it offers educational programs, many of which provide "close encounters" with the exhibits.

South Carolina

Ripley's Aquarium South Carolina, Myrtle Beach. It has both saltwater and freshwater exhibits, including a re-creation of the Amazon rain forest habitat.

South Carolina Aquarium, Charleston. Mountain streams flow through rivers, lakes, and salt marshes to the depths of the Atlantic Ocean. It has educational programs.

Tennessee

Tennessee Aquarium, Chattanooga. The world's largest freshwater aquarium, it features over nine thousand animals in natural habitats. It has special events and an IMAX theater.

Texas

Dallas Aquarium at Fair Park. It has both saltwater and freshwater exhibits and includes several rare desert fishes that have been bred at the aquarium.

Dallas World Aquarium and Zoological Garden. It has one of the largest freshwater fish exhibits in the world. The Web site has a live cam of manatees.

Moody Gardens, Galveston. The Rainforest Pyramid includes exhibits of fishes native to the rainforests of Africa, Asia, and the Americas.

Virginia

Chicoteague Island Aquarium. Experience local marine life up close. It has a "touch tank" with live horseshoe crabs, spider crabs, starfish, whelks, and more.

Virginia Marine Science Museum, Virginia Beach. It features aquatic animals and plants native to Virginia, from fresh water to the Chesapeake Bay to the depths of the Atlantic Ocean, with hands-on displays and educational programs as well as an IMAX theater.

Washington

Seattle Aquarium. Focused on marine exhibits, it offers a wide range of exhibits and educational programs, including summer camps.

References

General References

Bond, C. 1979. *Biology of fishes*. Philadelphia: W. B. Saunders.

Helfman, G. S., B. B. Collette, and D. E. Facey. 1997. *The diversity of fishes*. Malden, MA: Blackwell Science Publishers.

Lagler, K. F., J. E. Bardach, R. R. Miller, and D. R. Passino. 1977. *Ichthyology*. 2nd ed. New York: John Wiley & Sons.

Marshall, N. B. 1971. *Explorations in the life of fishes*. Cambridge, MA: Harvard University Press.

Moyle, P. B., and J. J. Cech. 1982. *Fishes: An introduction to ichthyology*. Englewood Cliffs, NJ: Prentice-Hall.

Reebs, S. 2001. *Fish behavior in the aquarium and in the wild*. Ithaca, NY: Cornell University Press.

Chapter One: Fish Basics

Depczynski, M., and D. R. Bellwood. 2005. Shortest recorded vertebrate lifespan found in a coral reef fish. *Current Biology* 15: R288–R289.

Everything Koi Web site. 2009. http://www.live-koifish.com/.

Holder, M. T., M. V. Erdmann, T. P. Wilcox, R. L. Caldwell, and D. M. Hillis. 1999. Two living species of Coelacanths? *Proceedings of the National Academy of Sciences* 96:12616–12620.

Nelson, J. S. 2006. *Fishes of the world*. 4th ed. Hoboken, NJ: John Wiley & Sons.

Pitcher, T. J., ed. 1986. *The behavior of Teleost fishes*. Baltimore, MD: Johns Hopkins University Press.

Ruber, L., M. Kottelat, H. Tan, P. Ng, and R. Britz. 2007. Evolution of miniaturization and the phylogenetic position of *Paedocypris*, comprising the world's smallest vertebrate. *BMC Evolutionary Biology* 7:38.

Chapter Two: Where Fishes Live

Cheng, C.-H. C., P. A. Cziko, and C. W. Evans. 2006. Non-hepatic origin of notothenioid antifreeze reveals pancreatic synthesis as common mechanism in polar fish freezing avoidance. *Proceedings of the National Academy of Sciences* 103:10491–10496.

Fast flying fish glides by ferry. 2009. BBC World News America Web site. http://news.bbc.co.uk/2/hi/science/nature/7410421.stm (accessed February 22, 2009).

Parashar, R. S., and T. K. Banerjee. 1999. Response of aerial respiratory organs of the air-breathing catfish *Heteropneustes fossilis* (Bloch) to extreme stress of desiccation. *Veterinarski Arhiv* 69:63–68.

U.S. Geological Survey Web site. 2009. http://wfrc.usgs.gov/research/aquatic%20ecology/STSaiki6.htm.

Van Wassenbergh, S., et al. 2006. Evolution: A catfish that can strike its prey on land. *Nature* 440:881.

Chapter Three: Fish Bodies

Arnegard, M., B. S. Jackson, and C. D. Hopkins. 2006. Time-domain signal divergence and discrimination without receptor modification in sympatric morphs of electric fishes. *Journal of Experimental Biology* 209:2182–2198.

Atema, J. 1971. Structures and functions of the sense of taste in the catfish (*Ictalurus natalis*). *Brain, Behavior, and Evolution* 4:273–294.

Bardach, J., J. H. Todd, and R. Crickmer. 1967. Orientation by taste in fish of the genus *Ictalurus*. *Science* 155:1276–1278.

Borowsky, R. 2008. Restoring sight in blind cavefish. *Current Biology* 18 (1): R23–R24.

Fine, M. L., J. P. Friel, D. McElroy, C. B. King, K. E. Loesser, and S. Newton. 1997. Pectoral spine locking and sound production in the channel catfish *Ictalurus punctatus*. *Copeia* 1997:777–790.

Gerlach, G., J. Atema, M. J. Kingsford, K. Black, and V. Miller-Sims. 2007. Smelling home can prevent dispersal of reef fish larvae. *Proceedings of the National Academy of Sciences* 104:858–863.

Mann, D. A., Z. Lu, and A. Popper. 1997. A clupeid fish can detect ultrasound. *Nature* 389:341.

Robison, B. H., and K. R. Reisenbichler. 2008. *Macropinna microstoma* and the paradox of its tubular eyes. *Copeia* 2008:780–784.

Sneddon, L., V. Braithwaite, and M. Gentle. 2003. Do fishes have nociceptors? Evidence for the evolution of a vertebrate sensory system. *Proceedings of the Royal Society, London, Series B* 270:1115–1121.

Yamamoto, Y., and W. R. Jeffery. 2000. Central role for the lens in cave fish eye degeneration. *Science* 289:631–633.

Chapter Four: Fish Lives

Biro, P. A., C. Beckmann, and J. A. Stamps. 2010. Small within-day increases in temperature affects boldness and alters personality in coral reef fish. *Proceedings of the Royal Society, London, Series B* 277:71–77.

Block, B. A., H. Dewar, S. Blackwell, T. Williams, E. Prince, C. Farwell, A. Boustany, S. Teo, A. Seitz, A. Walli, and D. Fudge. 2001. Migratory movements, depth preferences, and thermal biology of Atlantic bluefin tuna. *Science* 293:1310–1314.

Borawa, J. C., J. H. Kerby, T. Huish, and A. W. Mullis. 1979. Currituck Sound fish populations before and after infestation by Eurasian water-milfoil. *Proceedings of the Annual Conference of Southeastern Association of Fish and Wildlife Agencies* 32:520–528.

Brawley, S., and W. Adey. 1977. Territorial behavior of threespot damselfish (*Eupomacentrus planifrons*) increases reef algal biomass and productivity. *Environmental Biology of Fishes* 2:45–51.

Burkepile, D., and M. W. Hay. 2008. Herbivore species richness and feeding complementarity affect community structure and function on a coral reef. *Proceedings of the National Academy of Sciences* 105:16201–16206.

Cain, P., and S. Malwal. 2002. Landmark use and development of navigation behaviour in the weakly electric fish *Gnathonemus petersii* (Mormyridae; Teleostei). *Journal of Experimental Biology* 205:3915–3923.

Campbell, H. A., K. P. Fraser, C. M. Bishop, L. S. Peck, S. Egginton. 2008. Hibernation in an Antarctic fish: On ice for winter. *PLoS ONE* 3 (3): e1743. doi:10.1371/journal.pone.0001743.

Coleman, F., C. Koenig, K. M. Scanlon, S. Heppell, S. Heppell, and M. W. Miller. 2010. Benthic habitat modification through excavation by red grouper, Epinephelus morio, in the northeastern Gulf of Mexico. *Open Fish Science Journal* 3:1–15.

Dibble, E. D., K. J. Killgore, and S. L. Harrel. 1996. Assessment of fish–plant interactions. *American Fisheries Society Symposium* 16:357–372.

Elliot, J., and R. Mariscal. 1997. Acclimation or innate protection of anemonefish from sea anemones? *Copeia* 1997:284–289.

Fisher, H., B. M Wong, and G. Rosenthal. 2006. Alteration of the chemical environment disrupts communication in a freshwater fish. *Proceedings of the Biological Society* 273:1187–1193.

Glass, C. W., C. S. Wardle, and S. J. Gosden. 1992. A behavioural study of the principles underlying mesh penetration by fish. In *Fish Behavior in Relation to Fishing Operations,* edited by C. S. Wardle and C. Hollingworth, vol. 196, 92–97. ICES Marine Science Symposia. Bergen, Norway: International Council for the Exploration of the Sea.

Gomez-Laplaza, L. M., and E. Morgan. 2005. Time-place learning in the cichlid angelfish, *Pterophyllum scalare. Behavioral Processes* 70:177–181.

Killgore, K. J., R. P. Morgan II, and N. B. Rybicki. 1989. Distribution and abundance of fishes associated with submerged aquatic plants in the Potomac River. *North American Journal of Fisheries Management* 9:101–111.

Knight, T., M. McCoy, J. Chase, K. McCoy, and R. Holt. 2005. Trophic cascades across ecosystems. *Nature* 437:880–883.

Kohda, M. 2006. Territoriality of male cichlid fishes in Lake Tanganyika. *Ecology of Freshwater Fish* 4:180–184.

Lafaille, P., J.-M. Caraguel, and A. Legault. Temporal patterns in the upstream migration of European glass eels (*Anguilla anguilla*) at the Couesnon estuarine dam. *Estuarine and Coastal Shelf Science* 73:81–90.

Lohmann, K. J., N. F. Putman, and C.M.F. Lohmann. 2008. Geomagnetic imprinting: A unifying hypothesis of natal homing in salmon and sea turtles. *Proceedings of the National Academy of Sciences* 105:19096–19101.

Low, R. M. 1971. Interspecific territoriality in a Pomacentrid reef fish, *Pomacentrus flavicauda* Whitley. *Ecology* 52:648–654.

Lubbock, R. 1982. The clownfish anemone symbiosis: A problem of cellular recognition. *Parasitology* 82:159–173.

Mirza, R. S., W. W. Green, S. Connor, A. C. Weeks, C. M. Wood, and G. G. Pyle. 2008. Do you smell what I smell? Olfactory impairment in wild yellow perch from metal-contaminated waters. *Ecotoxicology and Environmental Safety* 72:677–683.

Munday, P. L., D. Dixon, J. Donelson, G. Jones, M. Pratchett, G. Devitsina, and K. Doving. 2009. Ocean acidification impairs olfactory discrimination and homing ability of a marine fish. *Proceedings of the National Academy of Sciences* 106:1848–1852.

Roopin, M., and N. Chadwick. 2009. Benefits to host sea anemones from ammonia contributions of resident anemonefish. *Journal of Experimental Marine Biology and Ecology* 370:27–34.

Todd, J. H., J. Atema, and J. E. Bardach. 1967. Chemical communication in social behavior of a fish, the yellow bullhead (*Ictalurus natalis*). *Science* 158:672–673.

Weis, J. S., G. Smith, T. Zhou, C. Bass, and P. Weis. 2001. Effects of contaminants on behavior: Biochemical mechanisms and ecological consequences. *BioScience* 51:209–218.

Westin, L. 1998. The spawning migration of European silver eel (*Anguilla anguilla* L.) with particular reference to stocked eel in the Baltic. *Fisheries Research* 38:257–270.

Yokogawa, T., W. Marin, J. Faraco, G. Pézeron, L. Appelbaum, J. F. Rosa, P. Mourrain, and E. Mignot. 2007. Characterization of sleep in zebrafish and insomnia in hypocretin receptor mutants. *PLoS Biology* 5 (10): e277.doi:10.1371/journal.pbio.0050277.

Chapter Five: Fish Reproduction

Basolo, A. 1990. Female preference for male sword length in the green swordtail, *Xiphophorus helleri* (Pisces: Poeciliidae). *Animal Behaviour* 40:332–338.

Chapman, D. D., M. S. Shivji. E. Louis, J. Sommer, H. Fletcher, and P. Prodöhl. 2007. Virgin birth in a hammerhead shark. *Biology Letters* 3:425–427.

Chapman, D. D., B. Firchau, and M. S. Shivji. 2008. Parthenogenesis in a large-bodied requiem shark, the blacktip *Carcharhinus limbatus*. *Journal of Fish Biology* 73:1473–1477.

The Earth Life Web. 2009. Fish Development Page. http://www.earth life.net/fish/development.html.

Flipbooks. 2009. The Zebrafish Flipbook Movie. http://www.bio .umass.edu/biology/karlstrom/ZFFlipbooks.html.

Florida Museum of Natural History Ichthyology Department. 2009. http://www.flmnh.ufl.edu/fish/organizations/ssg/sharknews/sn9/ shark9news8.htm.

Fricke, H., and S. Fricke. 1977. Monogamy and sex change by aggressive dominance in coral reef fish. *Nature* 266:830–832.

Friedman, M. 2008. The evolutionary origin of flatfish asymmetry. *Nature* 454:209–212.

Gilmore, R. G. 1993. Reproductive biology of lamnoid sharks. *Environmental Biology of Fishes* 38:95–114.

Gray, S. J., S. McKinnon, F. Y. Tantu, and L. M. Dill. 2008. Sneaky egg-eating in *Telmatherina sarasinorum*, an endemic fish from Sulawesi. *Journal of Fish Biology* 73:728–731.

Hamlett, W. C., A. M. Eulitt, R. Jarrell, and M. Kelly. Uterogestation and placentation in elasmobranches. *Journal of Experimental Zoology* 266:347–367.

Harrington, R. W. 1961. Oviparous hermaphroditic fish with internal self-fertilization. *Science* 134:1749–1750.

Johnson, D., J. Paxton, T. Sutton, T. Satoh, T. Saso, M. Nishida, and M. Miya. 2009. Deep-sea mystery solved: Astonishing larval transformations and extreme sexual dimorphism unite three fish families. *Biology Letters* 5:235–239.

Kimmell, C., W. Ballard, S. R. Kimmel, B. Ullman, and T. F. Schilling. 1995. Stages of embryonic development of the zebrafish. *Developmental Dynamics* 203:253–310.

Lamml, M., and B. Kramer. 2006. Differentiation of courtship songs in parapatric sibling species of dwarf stonebashers from southern Africa (Mormyridae, Teleostei). *Behaviour* 143:783–810.

Langerhans, R. B., C. A. Layman, and T. J. DeWitt. 2005. Male genital size reflects a trade-off between attracting mates and avoiding predation in two live-bearing fish species. *Proceedings of the National Academy of Sciences* 102:7618–7623.

Lorenz, K. 1952. *King Solomon's ring.* London: Methuen.

Plath, M., D. Blum, I. Schlupp, and R. Tiedemann. 2008. Audience effect alters mating preferences in a livebearing fish, the Atlantic molly, *Poecilia mexicana. Animal Behaviour* 75:21–29.

Ripley, J., and C. Foran. 2006. Differential parental nutrient allocation in two congeneric pipefish species (Syngnathidae: *Syngnathus* spp.). *Journal of Experimental Biology* 209:1112–1121.

Smith, C., and R. J. Wooten. 1995. The costs of parental care in teleost fishes. *Reviews of Fish Biology and Fisheries* 5:7–22.

Tinbergen, N. 1952. The curious behavior of the stickleback. *Scientific American* 187:22–26.

Warner, R. R. 1984. Mating behavior and hermaphroditism in coral reef fishes. *American Scientist* 72:128–136.

Chapter Six: Dangers and Defenses

Balirwa, J., et al. 2003. Biodiversity and fishery sustainability in the Lake Victoria basin: An unexpected marriage? *BioScience* 53:703–715.

Barry, D. 2008. On an infested river, battling invaders eye to eye. *New York Times.* http://www.nytimes.com/2008/09/15/us/15land .html?pagewanted=2&_r=1&ei=5070&emc=eta1 (accessed April 25, 2010).

Baumann, P. C. 1992. Methodological considerations for conducting tumor surveys of fishes. *Journal of Aquatic Ecosystem Stress and Recovery* 1:127–133.

———. 1998. Epizootics of cancer in fish associated with genotoxins in sediment and water. *Mutation Research* 411:227–233.

Berejikian, B., R. Smith, E. Tezak, S. Schroder, and C. Knudsen. 1999. Chemical alarm signals and complex hatchery rearing habitats affect antipredator behavior and survival of chinook salmon (*Oncorhynchus tshawytscha*) juveniles. *Canadian Journal of Fisheries and Aquatic Sciences* 56:830–838.

Breault, J. L. 1991. Candiru: Amazonian parasitic catfish. *Journal of Wilderness Medicine* 2:304–312.

Burkholder, J. M., H. B. Glasgow Jr., and C. W. Hobbs. 1995. Fish kills linked to a toxic ambush-predator dinoflagellate: Distribution and environmental conditions. *Marine Ecology Progress Series* 124:43–61.

Fishlore.com. 2009. Tropical Fish Information. http://www.fishlore.com/Disease.htm.

FlippersandFins.net. 2009. Fish Tank Granuloma. http://www.flippersandfins.net/FishTankGranuloma.htm.

Gisbert, E., and M. A. Lopez. 2007. First record of a population of the exotic mummichog, *Fundulus heteroclitus* (L., 1766) in the Mediterranean Sea basin (Ebro River delta). *Journal of Fish Biology* 71:1220–1224.

Glynn, P. W., I. C. Enochs, J. E. McCosker, and A. N. Graefe. 2008. First record of a pearlfish, *Carapus mourlani*, inhabiting the Aplysiid Opisthobranch Mollusc *Dolabella auricularia*. *Pacific Science* 62:593–601.

Gordon, S., D. Reznick, M. Kinnison, M. Bryant, D. Weese, K. Räsänen, N. Millar, and A. Hendry. 2009. Adaptive changes in life history and survival following a new guppy introduction. *American Naturalist* 174:34–45.

Greene, K. E., J. L. Zimmerman, R. W. Laney, and J. C. Thomas-Blate. 2009. *Atlantic Coast diadromous fish habitat: A review of utilization, threats, recommendations for conservation, and research needs.* Atlantic States Marine Fisheries Commission, Habitat Management Series #9. Washington, DC: Atlantic States Marine Fisheries Commission.

Invasive fish. 2009. USGS Great Lakes Science Center. http://www.glsc.usgs.gov/main.php?content=research_lamprey&title=Invasive%20Fish0&menu=research_invasive_fish.

Johnson, N. S., S. S. Yun, H. T. Thompson, C. O. Brant, and W. Li. 2009. A synthesized pheromone induces upstream movement in female sea lamprey and summons them into traps. *Proceedings of the National Academy of Sciences* 106:1021–1026.

Kitano J., D. Bolnick, D. Beauchamp, M. Mazur, S. Mori, T. Nakano, and C. Peichel. 2008. Reverse evolution of armor plates in three-spine stickleback. *Current Biology* 18:769–774.

Krkosek, M., M. A. Lewis, and J. P. Volpe. 2005. Transmission dynamics of parasitic sea lice from farm to wild salmon. *Proceedings of the Royal Society, London, Series B* 272:689–696.

Lafferty, K., and A. K. Morris. 1996. Altered behaviour of parasitized killifish increases susceptibility to predation by bird final hosts. *Ecology* 77:1390–1397.

National Oceanic and Atmospheric Administration (NOAA). 2007. http://www.noaanews.noaa.gov/stories2007/20071015_lionfish .html.

Rabalais, N., R. E. Turner, and W. Wiseman. 2002. Gulf of Mexico hypoxia, aka the "Dead Zone." *Annual Review of Ecology and Systematics* 33:235–263.

Samson, J. C., S. Shumway, and J. S. Weis. 2008. Effects of the toxic dinoflagellate *Alexandrium fundyense* on three species of larval fish: A food web approach. *Journal of Fish Biology* 72:168–188.

Sanderson, B. L., K. A. Barnas, and A. M. Rub. 2009. Nonindigenous species of the Pacific Northwest: An overlooked risk to endangered salmon? *BioScience* 59:245–256.

Santiago Bass, C., and J. S. Weis. 2009. Conspicuous behavior of *Fundulus heteroclitus* associated with high digenean metacercariae gill abundances. *Journal of Fish Biology* 74:763–772.

Sumpter, J. 1998. Xenoendocrine disruptors: Environmental impacts. *Toxicology Letters* 102–103:337–342.

Vogelbein, W. K., V. J. Lovko, J. D. Shields, K. S. Reece, P. Mason, L. W. Haas, and C. C. Walker. 2002. *Pfiesteria shumwayae* kills fish by micropredation, not exotoxin secretion. *Nature* 418:67–69.

Weis, J. S., G. Smith, T. Zhou, C. Bass, and P. Weis. 2001. Effects of contaminants on behavior: Biochemical mechanisms and ecological consequences. *BioScience* 51:209–218.

Chapter Seven: Watching Fishes

Baldwin. C. C., B. B. Collette, L. R. Parenti, D. G. Smith, and V. G. Springer. 1996. Collecting fishes. In *Diving for Science, "Methods and Techniques of Underwater Research,"* edited by M. A. Lang and C. C. Baldwin. Proceedings of the American Academy of Underwater

Sciences (Sixteenth Annual Scientific Diving Symposium). http://archive.rubicon-foundation.org/4678 (accessed June 28, 2009).

Cohen, C. L., and S. D. Schindler. 2000. *How many fish? (My first I can read book)*. New York: HarperCollins.

Cook, B., and C. Johnson. 2005. *The little fish that got away*. New York: HarperCollins.

Dakin, N. 1992. *The Macmillan book of the marine aquarium*. New York: Macmillan.

Pfister, M. 1993. *The rainbow fish*. New York: Scholastic Publications.

Seuss, Dr. 1960. *One fish, two fish, red fish, blue fish*. New York: Random House.

Smith, C. 1994. *Fish watching: An outdoor guide to freshwater fishes*. Ithaca, NY: Comstock/Cornell University Press.

Chapter Eight: Recreational Fishing

Angione, K. 2009. Texting your catch: New technology for recreational fishing data. *Coastwatch* (spring): 7–11. www.ncseagrant.org (accessed June 28, 2009).

Dying shad imitation. 2004. Ezknot.com. http://ezknot.com (accessed March 3, 2009).

Fobert, E., P. Meining, A. Colotelo, C. O'Connor, and S. J. Cooke. 2009. Cut the line or remove the hook? An evaluation of lethal and sublethal endpoints for deeply hooked bluegill. *Fisheries Research* 99:38–46.

Hemingway, E. 1952. *The old man and the sea*. New York: Charles Scribner's Sons.

Ice fishing: Ice fishing basics. New York State Department of Environmental Conservation. http://www.dec.ny.gov/outdoor/7733.html.

Inside Line. North American Fishing Club newsletter, January 28, 2009.

IUCN, the International Union for Conservation of Nature. 2009. http://www.iucn.org/?2213/Third-of-open-ocean-sharks-threatened-with-extinction (accessed April 28, 2010).

Maclean, N. 1992. *A river runs through it and other stories*. New York: Pocket Books (Simon & Schuster).

Quick glow in the dark fishing tackle. 2009. http://glowinc.com/glow-in-the-dark/fishing-lures.aspx (accessed March 3, 2009).

Ross, C. 2009. Casting for recovery throws lifeline to female cancer patients. *Bay Journal* 19:11.

———. 2009. Healing waters. *Bay Journal* 19:1.

Trigger X. 2009. Triggerx.com. http://www.triggerx.com/?promo8 (accessed March 3, 2009).

Walton, I. 1653. *The compleat angler; or, The contemplative man's recreation.* London: Richard Marriot.

Chapter Nine: Commercial Fishing

Araki, H., B. Cooper, and M. Blouin. 2009. Carry-over effect of captive breeding reduces reproductive fitness of wild-born descendants in the wild. *Biology Letters* 5:621–624.

Árnason, E., U. B. Hernandez, and K. Kristinsson. 2009. Intense habitat-specific fisheries-induced selection at the molecular pan I locus predicts imminent collapse of a major cod fishery. *PLoS ONE* 4 (5): e5529. doi:10.1371/journal.pone.0005529.

Conover, D. O., S. B. Munch, and S. A. Arnott. 2009. Reversal of evolutionary downsizing caused by selective harvest of large fish. *Proceedings of the Royal Society, London, Series B* 276:2015–2020.

Darimont, C. T., S. M. Carlson, M. T. Kinnison, P. C. Paquet, T. E. Reimchen, and C. C. Wilmers. 2009. Human predators outpace other agents of trait change in the wild. *Proceedings of the National Academy of Sciences* 106:952–954.

Diana, James. 2009. Aquaculture production and biodiversity conservation. *BioScience* 59:27–38.

Food and Agricultural Organization of the United Nations (FAO). 2007. *The state of world fisheries and aquaculture.* Rome: FAO.

Hay, M., and D. Burkpile. 2008. Herbivore species richness and feeding complementarity affect community structure and function on a coral reef. *Proceedings of the National Academy of Sciences* 105:16201–16206.

Jackson, J. B., et al. 2001. Historical overfishing and the recent collapse of coastal ecosystems. *Science* 293:629–638.

Kurlansky, M. 1997. *Cod: A biography of the fish that changed the world.* London: Random House.

Lynam, C., M. Gibbons, B. Axelsen, C. Sparks, J. Coetzee, B. Heywood, and A. Brierley. 2008. Jellyfish overtake fish in a heavily fished ecosystem. *Current Biology* 16:R492–R493.

McClenachan, L. 2009. Documenting loss of large trophy fish from the Florida Keys with historical photographs. *Conservation Biology* 23:636–643.

Mora, C., R. A. Myers, M. Coll, S. Libralato, T. Pitcher, et al. 2009. Management effectiveness of the world's marine fisheries. *PLoS Biology* 7 (6): e1000131. doi: 10.1371/journal.pbio.1000131.

Sadovy de Mitcheson, Y., A. Cornish, M. Domeier, P. L. Colin, M. Russell, and K. C. Lindeman. 2008. Reef fish spawning aggregations: A global baseline. *Conservation Biology* 22 (5):1233–1244.

Sainsbury, J. C. 1986. *Commercial fishing methods.* 3rd ed. Fishing News Books. Oxford: Blackwell Science.

U.S. Agency for International Development. 2006. *Fisheries opportunities assessment.* Narragansett, RI: Coastal Resources Center, University of Rhode Island, and Florida International University.

World Bank and Food and Agricultural Organization. 2008. *The sunken billions: Economic justification for fisheries reform.* Washington, DC: Agriculture and Rural Development Department, The World Bank.

Worm, B., N. Barbier, et al. 2006. Impacts of biodiversity loss on ocean ecosystem services. *Science* 314:787–790.

Worm, B., R. Hilborn, et al. 2009. Rebuilding global fisheries. *Science* 325:578–585.

Chapter Ten: Fish and Human Health

Bittman, M. Loving fish, this time with the fish in mind. *New York Times.* http://www.nytimes.com/2009/06/10/dining/10Seafood .html?emc=eta1 (accessed June 28, 2009).

Eating fish: Health benefits and risks. 2006. *JAMA, Journal of the American Medical Association.* http://jama.ama-assn.org/cgi/content/ full/296/15/1926 (accessed June 28, 2009).

Johnson, P. 2007. *Fish forever.* Hoboken, NJ: John Wiley & Sons.

Kraepiel, A. M., K. Keller, H. B. Chin, E. G. Malcolm, and F. M. Morel. 2003. Sources and variations of mercury in tuna. *Environmental Science and Technology* 37:5551–5558.

New York State Department of Health. 2009–2010. *Chemicals in sportfish and game: 2009–2010 health advisories.* http://www.health.state .ny.us/environmental/outdoors/Fish/Fish.htm.

Smackdown! EPA, FDA, and mercury in fish. 2009. Lab Notes Blog. Newsweek.com. http://blog.newsweek.com/blogs/labnotes/ archive/2009/04/24/smackdown-epa-fda-and-mercury-in-fish.aspx (accessed June 28, 2009).

Sunderland, E., D. P. Krabbenhoft, J. W. Moreau, S. A. Strode, and W. M. Landing. 2009. Mercury sources, distribution, and bioavailability in the North Pacific Ocean: Insights from data and models. *Global Biogeochemical Cycles* 23:1–14.

UC Berkeley Wellness Guide to Dietary Supplements: Cod Liver Oil. 2009. UC Berkeley wellness letter.com. http://www.wellnessletter .com/html/ds/dsCodLiverOil.php.

Yasumoto, T., D. Yasumura, M. Yotsu, T. Michishita, A. Endo, and Y. Kotaki. 1986. Bacterial production of tetrodotoxin and anhydro tetrodotoxin. *Agricultural and Biological Chemistry* 50:793–795.

Chapter Eleven: Research and Conservation

Costello, C., S. D. Gaines, and J. Lynham. 2008. Can catch shares prevent fisheries collapse? *Science* 321:1678–1681.

Franklin, H. B. 2007. *The most important fish in the sea*. Washington, DC: Island Press.

Humphries, P., and K. Winemiller. 2009. Historical impacts on river fauna, shifting baselines, and challenges for restoration. *BioScience* 59:673–684.

Jelks, H. L., et al. 2008. Conservation status of imperiled North American freshwater and diadromous fishes. *Fisheries* 33:372–407.

National Academies of Science. 2001. *Marine protected areas: Tools for sustaining ocean ecosystems*. Washington, DC: National Academies Press.

National Oceanic and Atmospheric Administration. 2009. NOAA's National Ocean Service Marine Protected Areas. http://oceanser vice.noaa.gov/topics/oceans/mpa/ (accessed April 25, 2010).

Planes, S., G. P. Jones, and S. Thorrold. 2009. Larval dispersal connects fish populations in a network of marine protected areas. *Proceedings of the National Academy of Sciences* 106:5693–5697.

Roberts, C. 2007. *The unnatural history of the sea*. Washington, DC: Island Press.

Silent streams? Escalating endangerment for North American freshwater fish: Nearly 40 percent now at-risk. 2008. U.S. Geological Survey. http://www.usgs.gov/newsroom/article .asp?ID=2019&from=rss_home (accessed April 25, 2010).

White, R. M., A. Sessa, C. Burke, et al. 2008. Transparent adult zebrafish as a tool for in vivo transplantation analysis. *Cell Stem Cell* 2:183–189.

Index

aquariums, 9, 18, 28, 37, 60, 63,
    64, 66, 73, 91, 93–95, 97, 108,
    111, 114–119, 123, 155, 157, 164,
    170, 181, 183–187, 189, 197
archerfish, 34, *35*, 51
Arctic, 6, 22, 149
armor, 2, 3, 29, 100, 195
Árnason, Einar, 143, 198
Arnegard, Matt, 45, 190
Ash Meadows, 21
Astyanax mexicanus, 36, *Plate C*
Atheriniformes, 5
Atlantic Ocean, 24, 40, 50, 56, 58,
    67, 78, 82, 83, 103, 111, 138,
    141–144, 147, 148, 152–154, 159,
    162, 164, 165, 173, 174, 179,
    186, 187, 191, 194, 195
Atlantic States Marine Fisheries
    Commission, 103, 195

backbone. *See* spine
Balirwa, John, 108, 194
Baltic Sea, 57, 193
barb, 116; rosy, 116
barbel, 40, 41, 52
Bardach, John, 40, 189, 190, 192
Barnum, P. T., 117
barracuda, 5, 23, 52, 68, *106*, 162
barreleye fish, 37
bass, 5, 14, 16, 64, 131, 152; large-
    mouth, 109, 164; rock, 109;
    sea, black, 157, 165, 166; sea,
    Chilean, 141, 164, 165, 181;
    smallmouth, 110, 109; striped,
    13, 14, 56, 103, 152, 153, 159,
    163, 165
Bathylynchnops, 35
bathypelagic zone, 20
Baykeeper organizations, 180
Bellwood, David, 9, 189
Belonidae, 107
Beloniformes, 5
Benchley, Peter, 105, 126, 130

Benguela current, 145
benthic zone, 18, 19, 23, 50, 51, 98,
    139, 178, 191
Betta splendens, 62, 87, *Plate F*
billfish, 107, 128
bioaccumulatation, 104, 163
biodiversity, 50, 157, 179, 194, 198,
    199
bioluminescence, 20, 36, 37, 46,
    59, 76, 100, *Plate B*
biomagnification, 104
birds, 3, 11, 33, 52, 66, 96, 97, 99,
    103, 145–147, 152, 185, 196
bitterling, 81
Black River, 95
Black Sea, 173
blastema, 102
bleaching, coral, 105
blennies, 5, 18, 68
Block, Barbara, 54, 191
blobfish, 19
blood, 2, 12, 17, 22, 32–34, 45, 65,
    86, 88, 93, 95, 97–99, 107, 126,
    132, 162, 167, 171, 172
bluefish, 52, 141, 152, 159, 163, 165
bluegill, 128, 133, 197
Blue Ocean Institute, 165
Bluin, Michael, 155
bogue, 11
bone, 1, 4, 5, 7, 10, 26, 27, 30, 39,
    40, 47, 160, 167, 174
bony fishes, 2–6, 9, 11, 13, 29, 32,
    39, 47, 83, 100
Boops boops, 11
Borawa, J. C., 71, 191
Borowsky, Richard, 36, 190,
    Plate C
Bothus lunatus, *Plate D*
bowfin, 4
brackish water, 115, 163
branzini, 157
Brawley, Susan, 62, 191
bream, 24

174; devil, 174; manta, 54; sting,
84, 107; torpedo, 44, 107
redds, 79
Red Sea, 24, 119
red tide, 103
reduction fishery, 152
reef: artificial, 24; coral, 9, 11, 23,
24, 38, 41, 43, 49, 50, 57, 58, 62,
64, 68, 69, 90, 103, 105, 107,
112, 114, 116–119, 123, 142, 146,
163, 172, 178, 179, 181, 186,
189–194, 198, *Plate B*
Regalecus glesne, 11
regeneration, 30, 102, 171
Reisenbichler, Kim, 37, 190
Reisman, Howard, 168
remora, 67, *68*
remotely operated vehicle
(ROV), 7
respiration, 17, 32, 88, 94, 115, 190
rete mirabile, 33
retina, 34, 35, 37
Rhincodon typus, 10, 50, *Plate A*
Rhinecanthus rectangulus, 11,
Plate A
Rhode Island, 48, 111, 199
Rhodeus amarus, 81
Ripley, Jennifer, 86, 194
Riverkeeper organizations, 180
rivers, 7, 13, 14, 38, 44, 55–58, 65,
69, 71, 95, 98, 100, 102, 103,
109, 110, 112, 118, 119, 122,
126, 127, 134, 171, 173–175,
183–186, 192, 193, 195
Rivulus marmoratus, 92
roach, 42
Roberts, Callum, 174, 176, 200
Robison, Bruce, 37, 190
rockfish, 5, 19, 64, 165, *Plate B*
rockhoppers, 135, *136*, 146
Roopin, Modi, 70, 192
rosefish, 24
roundworm, 96, 163, 170

rust, 94
Rutilus rutilus, 42

sablefish, 141, 165
Sadovy, Yvonne, 142, 198
salamander, 99
Salmo salar, 154
salmon/Salmonidae, 13, 14, 31,
38, 40, 54–58, 61, 79, 83, 97,
102, 109, 110, 125, 128, 134,
138, 141, 152, 154–156, 158,
161, 163, 165, 174, 181, 192, 195,
196, *Plate C, Plate G*; Atlantic,
154, 174; Chinook, 55, 195;
chum, 55; coho, 157; hump-
back, 55; Pacific, 152, 163, 172;
pink, 138, *Plate G*; sockeye, 138,
Plate C
Salmoniformes, 5
salt marsh, 71, 72, 112, 186, 201
Samson, Jennifer, 104, 196
Sanderson, Beth, 109, 196
Sander vitreus, 138
Santiago Bass, Celine, 97, 196
Saprolegnia, 94
sardine, 54, 55, 138, 141, 145, 158,
175, 176; Pacific, 138
sardine run, 54, *55*, 176
Sardinops sagax, 54, 138
Sargasso Sea, 56, 72
Sargassum fish, 72
Sarotherodon galilaeus, 121
sashimi, 163
Sass, Greg, 109
scales, 2, 9, 10, *29*, 30, 33, 42,
43, 89, 94, 161; ctenoid, 9,
29; cycloid, 9, *29*; ganoid, *29*;
placoid, *29*
Schindler, S. D., 119, 196
schooling, 2, 25, 30, 31, 42, 43, 52,
58–60, 64, 88, 101, 104, 135,
139, 147, 176, *Plate B, Plate E*
Scorpaeniformes, 5

About the Author

Judith S. Weis is a professor of biology at Rutgers University in Newark and has been studying fishes for most of her career. Most of her research work has been in northern New Jersey, but she has also studied estuaries and salt marshes from Massachusetts to Florida and fishes in mangrove swamps in Indonesia and Madagascar. She is particularly interested in fish development, behavior, and feeding ecology and how they cope with stresses such as pollution, invasive species, and parasites. She is a Fellow of the American Association for the Advancement of Science (AAAS) and has served on advisory committees to the National Academy of Sciences, National Oceanic and Atmospheric Administration (NOAA), and the Environmental Protection Agency. She has been a Fulbright Senior Specialist in Indonesia and was the president of the American Institute of Biological Sciences (AIBS) in 2001. She has published over two hundred scientific research articles and the book *Salt Marshes: A Natural and Unnatural History*, co-authored by Carol A. Butler (Rutgers University Press, 2009).